Sonic Phantoms

Sonic Phantoms

Composition with Auditory Phantasmatic Presence

Barbara Ellison and Thomas Bey William Bailey

Edited by
Francisco López

BLOOMSBURY ACADEMIC
NEW YORK · LONDON · OXFORD · NEW DELHI · SYDNEY

BLOOMSBURY ACADEMIC
Bloomsbury Publishing Inc
1385 Broadway, New York, NY 10018, USA
50 Bedford Square, London, WC1B 3DP, UK
29 Earlsfort Terrace, Dublin 2, Ireland

BLOOMSBURY, BLOOMSBURY ACADEMIC and the Diana logo
are trademarks of Bloomsbury Publishing Plc

First published in the United States of America 2020
This paperback edition published in 2022

Copyright © Barbara Ellison, Thomas Bey William Bailey and Francisco López, 2020

Cover design: Louise Dugdale
Cover image © Barbara Ellison

All rights reserved. No part of this publication may be reproduced or transmitted in any form or by any means, electronic or mechanical, including photocopying, recording, or any information storage or retrieval system, without prior permission in writing from the publishers.

Bloomsbury Publishing Inc does not have any control over, or responsibility for, any third-party websites referred to or in this book. All internet addresses given in this book were correct at the time of going to press. The author and publisher regret any inconvenience caused if addresses have changed or sites have ceased to exist, but can accept no responsibility for any such changes.

A catalog record of this book is available from the Library of Congress

ISBN: HB: 978-1-5013-4702-3
PB: 978-1-5013-9176-7
ePDF: 978-1-5013-4704-7
eBook: 978-1-5013-4703-0

Typeset by Integra Software Services Pvt. Ltd.

To find out more about our authors and books visit www.bloomsbury.com and sign up for our newsletters.

Ye Highlands and ye Lowlands,
Oh, where hae ye been?
They hae slain the Earl Amurray,
And Lady Mondegreen.

—Sylvia Wright

For Francisco
—*B. E.*

For my companion animals past and present: Fritz, Lily, Thor, Kira
—T. B. W. B.

Contents

List of Figures	xiii
Preface	xv
Author's Note	xviii
Acknowledgments	xix

1 Phantasmagenics — 3
  1.1 Sonic Phantoms as Emergent Presence — 3
  1.2 Presence by Apophenia — 11
  1.3 Apophenia and Creativity — 22
  1.4 Creativity with Ambiguity — 31
  1.5 Ambiguity as Intentional Practice — 38

2 Inducing the Phantasmatic — 45
  2.1 Compositional Phantasmatic Strategies and Techniques — 45
    2.1.1 Repetition — 45
    2.1.2 Persistence — 55
    2.1.3 Layering — 65
    2.1.4 Noise — 74
    2.1.5 Accentuation — 79
  2.2 Realms of Sonic Phantasmatic Experience — 80

3 *Phantasma Instrumentalis*: The Realm of the Instrument
  Compositional Series: *Harp Phantoms* — 85
  3.1 Sonic Exploration of the Instrument — 85
  3.2 Instrument Preparation and Sonic Blotscapes — 87
  3.3 Ritualization in Performance — 91
  3.4 Structure and Possible Variations — 93
  3.5 Inherent Patterns — 94
  3.6 Auditory Streaming — 96
  3.7 Sonic Figure and Ground — 100
  3.8 Listening Modes and Perceptual Competition — 101
  3.9 Accentuation of Harp Sonic Phantoms — 102
  3.10 A Second Life of *Harp Phantoms* in the Studio — 104

4	*Phantasma Materialis*: The Realm of the Object	
	Compositional Series: *Drawing Phantoms*	107
	4.1 Initial Material Explorations (for a Literal 'Phono-Graphic' Performance)	107
	4.2 *The Drawing Room*	109
	4.3 Transcendent 'Boundary Loss'	112
	4.4 Automatic Drawing—Ghosts and Dissociation	113
	4.5 *Drawing Phantoms* Performance	120
	4.6 Expanding by Narrowing—Everyday and Induced Trance	123
	4.7 Loop Multiplicity	130
	4.8 *Drawing Phantoms* Séance	131
	4.9 'EVP' and 'OVP'	135
5	*Phantasma Humana*: The Realm of the Voice	
	Compositional Series: *Vocal Phantoms*	147
	5.1 *Katajjaq*	147
	5.2 *Vocal Phantoms #18* (for Live Voices)	151
	5.3 Ancient and Recent Vocal Techniques of Illusion	153
	5.4 *CyberSongs*—Text-to-Speech-to-Song Pieces	158
	5.5 Micro-Temporal Mechanisms	162
	5.6 Semantic Satiation and Semantization	167
6	*Phantasma Naturalis*: The Realm of Nature	
	Compositional Series: *Natural Phantoms*	177
	6.1 Listening and Recording	177
	6.2 Natural Polyphony	182
	6.2.1 Layering	183
	6.2.2 Interlocking	184
	6.2.3 Transitions	189
	6.3 *Natural Phantoms*	191
	6.4 Beyond Composition	192
7	*Coda*	
	Otoacoustic Emissions: The Phantom Within	195
Bibliography		200
Index		213

Figures

1.1	*The Cloud That Looks Like Ireland*	2
1.2	*Phantom Portrait*	12
1.3	*The Dead Pigeon on a Branch That Looks Like Ireland* and *The Brillopad That Looks Like Ireland*	15
1.4	The 'Necker cube' and 'Rabbit and Duck'	18
1.5	*Suède paréidolie*	21
1.6	Bistable pareidolic images	23
1.7	'White's illusion'	37
1.8	Examples of some of my own digitally generated 'blotscapes'	42
1.9	*Something Bumped against a Wall at Work and Made a Painting of a Snowy Town*	43
1.10	*Snowy Town*	43
2.1	*The Third Voice*	44
3.1	*Harp Phantoms* compositional series	84
3.2	A workshop session experimenting with different materials	87
3.3	Diagram and photographic illustrations with harp 'preparation'	89
3.4	Part of the score for the compositional series *Harp Phantoms*	93
4.1	*Drawing Phantoms* séance	106
4.2	A selection of prototypical 'traces' of the drawing process from the collective performance *The Drawing Room*	110
4.3	*The Drawing Room* collective performance	111
4.4	Graphic results from some of my personal 'automated drawing' sessions	119
4.5	Live performances *Drawing EVP* and *Drawing OVP*	122
4.6	*Drawing Phantoms* séance	132
4.7	Raising the noise floor	143
5.1	Iqaluit, Baffin Island, Nunavut, Canada	146
5.2	Section of the score/diagram for live performance set-up of *Vocal Phantoms #18*	153
5.3	Live performance with real-time video projection of *CyberSongs*	160
5.4	An avatar 'performs' (via Text-to-Speech-to-Song) a sound installation version with video projection of some of my *Vocal Phantoms*	173

6.1	Fieldwork location for environmental recordings and on-site listening at the Great Otway National Park, Australia	176
6.2	Fieldwork location for environmental recordings and on-site listening in the rainforest of Borneo, Malaysia	179
6.3	Interlocking patterns of frog and insect sounds as shown in a spectrogram (top image) from a prototypical example of the nocturnal sonic environment of the Cardamom Mountains rainforest in Cambodia (bottom image)	185
6.4	Fieldwork locations for environmental recordings and on-site listening in the savanna environment of Mmabolela Reserve, northeastern South Africa (top image). Spectrogram from a prototypical nocturnal recording (bottom image)	186
6.5	Spectrograms with prototypical examples of micro-timbral structure in several isolated individual animal calls	188
6.6	Two spectrograms showing a prototypical example of the multilayered rainforest dusk transition (from Borneo, Malaysia) at different scales	190
7.1	The Phantom Within	194

Preface

The story that is about to unfold is, at its simplest, the story of a particular method of experimentation with sound, and the unique perceptual, aesthetic, and experiential territories to which it grants us access. By extension, then, this is a story of 'experimental' sonic practice or music making. As it has been for decades now, we do not go forward with the assumption that readers have a monolithic understanding of that term. Many who voluntarily claim 'experimental music' as their area of special expertise are using that term to validate a set of actions and impressions that may be mildly unusual, and many others are involved with the comparably more rigorous activity of moving beyond the *symbolic* use of art towards an *indexical* use of it (as per Morag Josephine Grant's suggestion).[1]

The difference between the two is, essentially, the difference between the concept of art as a simple representative mirror reflecting the experiential world back to us, and a concept that sees art from a more radically transformative perspective, one that sees the representation of the world 'as it actually is,' not as a confirmation of closure, but only as a point of departure; an invitation to augment that world in ways hitherto impossible or unimaginable. Of course, these functions do not have to be mutually exclusive (mirror images can often be catalysts for either subtle or sweeping changes on all levels of human interaction). Yet it is the latter function that this work is most intimately connected with, and in our attempt to contribute to this creative lineage, there is a golden thread running through this work—one that seems essential for creating personalized languages of creative experimentation—relating to how perception itself is an active rather than passive process.

The vast realm of sonic experimentation from at least the mid-point of the twentieth century has been intensely involved with teaching this lesson; encouraging numerous variations on the theme of 'listening to listening.' This litany of achievements will not be recited again at this early stage, but suffice it to say for the moment that this simple reconfiguring of the listener as a sound *producer or creator* has inspired significant inquiries into fields from the technological to the spiritual, and has therefore had implications for the hybridization of culture that go well beyond the bounds of musical or audial practice.

We submit that one of the key realizations provided upon rejecting the 'passive' nature of audio experience is the degree to which we, as humans, ascribe meaning to the apparently meaningless. This is not an activity that has abated with the advancement of techno-scientific knowledge; in fact, the torrential downpour of digitized data that we now encounter is every bit as defined by meaningless 'noise' as the features of

---

[1] Grant, "Experimental Music Semiotics."

some primeval forest. In all these cases, the attempts at subjective 'sculpting' of this noise into meaningful information have been as much an aesthetic exercise as an evolutionary strategy. Many of the thinkers cited in this book would argue, in fact, that creativity *is* the discernment of meaning within chance or random configurations of perceptible data; that drawing outlines around recognizable 'figures' inhabiting a more chaotic 'ground' of seemingly unrelated data fragments is the perennial act of artistic imagination. That the reality arising from such interpretations is not objectively true is of no consequence if, again, our purpose is not simply to acknowledge the mirror image but to distort it into something more suitable to our respective visions of an ideal world.

As we dive below the turbulent surface waters of apparent meaninglessness and related phenomena, we will unveil many of the techniques that are used to extract intensely personal meaning, from the simple act of intense repetition to more physically and psychically demanding actions. We will also be confronted with an astonishingly diverse set of actors and activities, whose multiplicity strongly suggests the universal character of our pattern-seeking behavior and our desire to become co-creators by means of this behavior. Certainly, representatives of historically 'avant-garde' movements from text-sound poetry to electroacoustic composition have made use of this type of creativity with aesthetic or psycho-social aims in mind. However, the institutionally recognized portions of that avant-garde are far from the only culture to adapt this strange ambiguity to their needs, and their research far from being the only manifestation of such creativity.

In folk traditions such as the Inuit *katajjaq* singing, it becomes the centerpiece of an often comical challenge game and leaks into a host of daily functions, from merely alleviating boredom to calming restless infants. Elsewhere, far away from the realm of 'art for art's sake,' there is the paranormal/esoteric subculture of EVP (electronic voice phenomena) recordists, for whom the 'figure' arising out of the 'ground' of indistinct audio is something to be hunted out and then used almost as a sort of divination tool, if not interpreted as an actual entity capable of communication. Equally intriguing are the ways in which our perceptual apparatus responds to the sonic environments that exist free of human agency: the natural world is rife with special kinds of organized polyphony, astonishing in their precision and clarity, which can produce the sort of immersive or entrancing effect common to the most deliberately complex experimental music.

The significance of all of the above will be explored over the course of this book, illustrated with a wide variety of sonic and compositional creative undertakings, prominently those from my personal practice, as they were the actual genesis of this book (these can be listened to in the companion website to this book at http://www.sonicphantoms.com).

It is our hope that these investigations into art-historical achievements and manifestations, along with explorations of sonic environments and audio cultures spanning the globe, will merge with the detailed records of personal experimentation to provide an understanding of the term I propose for the aforementioned creativity in audible form: 'sonic phantoms.' This term, which I have used to describe an ongoing series of sound compositions developed so far over the past decade (2009–2019), denotes a unique type of perceptual auditory illusion differing from other commonly cited illusions in a salient respect: namely, its profound potential to display the full extent of our 'activeness' as listeners and the creative roles that we can play even when we are not acknowledged as the primary actors or composers that we should perhaps necessarily be.

Author's Note

Although this is a two-author book, all the original compositional work included has been conceived and created solely by Barbara Ellison. Therefore, all the first-person mentions in the text ('I,' 'my,' 'me,' etc., which always relate to these compositional perspectives and developments) refer to Barbara Ellison.

Acknowledgments

I wish to sincerely thank my creative collaborators along this journey, in particular Angélica V. Salvi, Rhodri Davies, and the Trickster collective—Nathalie Smoor, Nina Boas, Ieke Trinks, and Marielle Verdijk—to all of whom I am forever grateful for their generous time, creativity, and inspirational phantom-producing talents. I would also like to acknowledge and extend my heartfelt thanks to the organizers, staff, hosts, and collaborators of the following projects and events in which I participated and which had a decisive influence in the multiple compositional outcomes described in this book:

- The Mamori Sound Project workshop and residency, Mamori lake, Amazon, Brazil.
- The Danum Valley Field Centre (DVFC) in Sabah, Borneo.
- Staff and hosts of our Sonic Mmabolela workshop and residency at Mmabolela reserve, Limpopo, South Africa (with special mention to Neil Lowe, Mark and Lesley Berry, James Webb and his family, William (Bolpetse) and Maria (Khani) and their family, Madikana, and particularly Frans (Mabig), who is still with us in spirit).
- Stephanie Pan and Stelios Manousakis for our expedition to Iqaluit, Baffin Island; our amazing hosts there—Vinnie, Hallauk, Ian, and Tank Karetak; our superb throat singers—Sandra Ikkidluak and Crystal Mullin; and Dr. Mary Piercey-Lewis, music teacher at Inuksuk High School, Iqaluit.
- Many thanks also go to Dr. Joseph Jordania and Prof. Rusudan Tsurtsumia for the invitation to present my research at the Sixth International Symposium for Polyphonic Music in Tbilisi in 2012. Special thanks as well to our legendary hosts *extraordinaires* Vakho and Islam Pilpani (and their entire family), for the chance of a lifetime to immerse ourselves in Georgian polyphony in their warm and welcoming home in Svaneti. Islam sadly passed away in 2017, but his legacy is extraordinary and his inspiration very much lives on through his family, friends, and fellow musicians from all around the word.

I received my PhD from the University of Huddersfield in 2014 for my research and compositional work into 'Sonic Phantoms.' I owe special thanks to Liza Lim for her invaluable support and encouragement and continued friendship and would also like to sincerely thank Deborah Middleton, Bryn Harrison, and the inspirational and very sadly missed Bob Gilmore. My guardian angel Henriëtte Vluggen-Hamaekers from the Prince Bernhard Cultuurfonds also deserves a special mention, for her splendid assistance during those PhD years.

I want to kindly thank Leah-Babb Rosenfeld and Ally Jane Grossan from Bloomsbury for their faith, interest, and willingness to take on this project with such an unknown as myself. A special mention here also to Deborah Maloney for her meticulous, efficient and cheery contribution to managing and preparing our final text for publication. We are very pleased to be part of the Bloomsbury Sound Studies series. This book was a total independent venture written with zero financial support (despite my great efforts to secure some). Bearing this in mind, and even though he is co-author of this volume, I wish to personally acknowledge what a rare lucky find Thomas Bey William Bailey was for this book. He knew exactly how to work and play with such material, with a true 'trickster' spirit, and was a joy to collaborate with.

A prolonged creative and emotional journey like this would have never been possible without the support of my truly wonderful and beloved family, to whom I am forever grateful. My beautiful mum, who is missed so much, would have been most proud. My ever-loving dad will definitely get a copy and may even give reading it a go, albeit no small challenge.

Finally, I have had the great fortune to share my *Sonic Phantoms* journey in all its forms with my partner Francisco López, and it is to him that I owe the greatest debt of all. His phantasmatic music is loaded with the most intriguing and powerful sonic phantoms and his support, advice, feedback, and inspirational insight, bestowed with such a generous and playful spirit throughout this project, was completely priceless; without this, this book would never have happened at all. I dedicate this book wholeheartedly to him.

—B. E.

The curious traveller, who, from open day,
Hath passed with torches into some huge cave,
The Grotto of Antiparos, or the Den
In old time haunted by that Danish Witch,
Yordas; he looks around and sees the vault
Widening on all sides; sees, or thinks he sees,
Erelong, the massy roof above his head,
That instantly unsettles and recedes,—
Substance and shadow, light and darkness, all
Commingled, making up a canopy
Of shapes and forms and tendencies to shape
That shift and vanish, change and interchange
Like spectres,—ferment silent and sublime!
That after a short space works less and less,
Till, every effort, every motion gone,
The scene before him stands in perfect view
Exposed, and lifeless as a written book!—
But let him pause awhile, and look again,
And a new quickening shall succeed, at first
Beginning timidly, then creeping fast,
Till the whole cave, so late a senseless mass,
Busies the eye with images and forms
Boldly assembled,—here is shadowed forth
From the projections, wrinkles, cavities,
A variegated landscape,—there the shape
Of some gigantic warrior clad in mail,
The ghostly semblance of a hooded monk.
Veiled nun, or pilgrim resting on his staff:
Strange congregation! yet not slow to meet
Eyes that perceive through minds that can inspire.

—William Wordsworth (*The Prelude*, Book VIII, 1850)[1]

---

[1] This beautiful apophenic poem of Wordsworth's was used as an example by Matteo Meschiari in "support of its universality,... not so much as a literary curiosity, but rather for the extraordinary exactitude in the description of apophenic dynamics." Meschiari, "Roots of the Savage Mind"; Wordsworth et al., *The Poetical Works of William Wordsworth*.

Figure 1.1 *The Cloud That Looks Like Ireland*, a grand example of pareidolia. Photograph by Lynn Donovan-Witt with kind permission also by Broadsheet.ie, publishers of the online and printed source image in the book *The Broadsheet Book of Unspecified Things That Look Like Ireland*, edited by Aidan Coughlan, 2013.

1

Phantasmagenics

The environment as we perceive it is our invention.

—Heinz von Foerster

1.1 Sonic Phantoms as Emergent Presence

The world that we experience is never more than a step removed from becoming something fascinatingly unfamiliar. Whole libraries could be formed from the available anecdotal accounts dealing with our perceptive abilities misreading the information they are presented with, and it could also be said that limitless mutations of art and cultural life have sprung forth from this fertile ground. To introduce this concept, the novelist Witold Gombrowicz's amusing account of a New Year's Eve dance party, in which "it was not the music that was eliciting the dance ... but the dance that was drawing forth the music," will serve our purposes here as well as any.[2] Though the novelist warns the reader that he may have been influenced in this misperception by "a turkey, quite a bit of vodka and wine," Gombrowicz makes numerous allusions to the mundane atmosphere that framed this revelation (e.g., "the dance of overworked, ordinary everydayness, kicking up its holiday heels, the dance of drabness ..."). In short, his was not an exceptional state of mind, nor would it be so for an individual not possessed of his celebratedly unique perspective. If anything, his status as an acclaimed man of letters should have made him less likely to be ambushed by his own senses to the point where such an error would be possible.

Philosopher Andy Clark[3] has described how human brains (whether belonging to 'acclaimed men of letters' or not) evolutionarily became master organs of prediction in order to deal with our incessantly changing, noisy, and ambiguous environment in an efficient and effective manner:

> Brains like ours, this picture suggests, are predictive engines, constantly trying to guess at the structure and shape of the incoming sensory array. Such brains are incessantly pro-active, restlessly seeking to generate the sensory data for themselves using the incoming signal (in a surprising inversion of much traditional wisdom) mostly as a means of checking and correcting their best top-down guessing.[4]

---

[2] Gombrowicz, *Diary, Volume 1*, 63.
[3] Andy Clark is Professor of Cognitive Philosophy at the University of Sussex.
[4] Clark, *Surfing Uncertainty*, 3.

The word 'predictive' is the key here, as this relates to a fundamental 'trick' or strategy that the brain uses to 'experience' and make sense of perceptual noise and ambiguity. This concept of the brain being a predictive engine is not novel, and can be traced back to a millennium ago by Ḥasan Ibn al-Haytham (an Arab mathematician, astronomer, and physicist of the Islamic Golden Age) who wrote, "not everything perceived by the sense of sight is perceived by pure sensation; rather many visible properties are perceived by judgement and inference."[5] More recently on the historical timeline, Jakob Hohwy referred to "a distinct Kantian element" to this idea, namely that "perception arises as the brain uses its prior conceptions of the world to organize the chaotic sensory manifold confronting the sensory system."[6] Immanuel Kant suggested that the mind used innate mental concepts to make sense of our sensory world and that there was a clear distinction between sensory experiences —or 'appearances'—and 'things in themselves.'[7] Following from Kant, it was Hermann von Helmholtz who first grasped the fundamental idea of the brain as a hypothesis tester. Helmholtz's idea of 'unconscious inference' was that the brain uses learned predictions to *infer* the causes of incoming sensory data, a sort of informed guesswork.[8] We will see over time that these core ideas, such as inference, prediction, expectation, and attention, in various ways are foundational to our exploration of sonic phantoms.

The above concepts have been codified into a framework or paradigm known as predictive processing (PP) or predictive coding (PC), which has become applicable to fields as diverse as computational and cognitive neuroscience, philosophy, psychology, psychiatry, robotics, and artificial intelligence. In simple terms, PP posits that the brain infers what is 'out there' by continually *predicting* what is out there, and then upgrades those predictions as it goes along. The brain predicts incoming sensory data about what can be perceived both from within and from outside of the body. In this PP framework, the term 'prediction' is used not in the usual way of us anticipating future events (conscious guessing), but as a different kind of 'guessing.' As Clark puts it, it is an "automatically deployed, deeply probabilistic, non-conscious guessing that occurs as part of the complex neural processing routines that underpin and unify perception and action."[9] The theory proposes that the brain constructs a hierarchical *generative* model of the world, a 'top-down' (knowledge-driven) model, which it updates if necessary, according to the 'bottom-up' incoming stimuli that it tries to fit its generated models by predicting the incoming signal. It is important to emphasize how different the PP perspective is from the more dominant 'classical' accounts of a passive form of perception that, up until recently, were the most commonly accepted

---

[5] Ibn al-Haytham, *The Optics of Ibn Al-Haytham*, 11.3.26.
[6] Hohwy, *The Predictive Mind*, 5.
[7] Swanson, "The Predictive Processing Paradigm has Roots in Kant."
[8] This process can be formalized as Bayesian inference whereby a probabilistic prediction or prior is combined with observed sensory data (likelihood) to compute a posterior probability (posterior).
[9] This account makes a strong connection to the 'Bayesian Brain Hypothesis' (more on this later) and outlines a vision of the brain as an engine of multilevel probabilistic prediction. This is a unifying set of ideas known as 'Predictive Processing' models or PP framework, directly related to the 'Predictive Coding' theory and work of Karl Friston and others. See Friston, "The Free-Energy Principle."

way of understanding this subject. Formerly, perception was thought of as being a simple 'feed-forward' process, in which the standard picture was one of a passive brain functioning primarily as a feature detector, being stimulus driven, with perception working from mainly the bottom-up and from the outside-in. In this traditional view, sensory information makes its way from the outside in (e.g., through the eyes or ears) and makes its way up, reaching higher and higher levels of processing until the percept is 'understood' by the brain.

Predictive processing turns this traditional picture of perception completely on its head and in direct opposition—it shows a highly 'active' view of the brain and that perception is an active constructive process. Perceptual phantoms, sonic phantoms, or illusions reveal just how much perception comes as much from the inside-out (top-down) as from the outside-in (bottom-up). Our experience of sensory illusions provides a constant reminder as to the ease of deceiving our own perceptual systems. Illusions are commonly thought of as the perceptual oddities that deviate from our normal, objective, properly working perceptual experience. However, the fact of the matter is that we and our perceptual systems are never directly experiencing an external, objective reality. As Anil Seth elegantly puts it, "All our perceptions are active constructions, brain-based best guesses at the nature of a world that is forever obscured behind a sensory veil. Visual illusions are fractures in the Matrix, fleeting glimpses into this deeper truth."[10] What is seen or heard is never just the direct sensory signal from the eyes or ears, but a combination of those signals with what our brains are *expecting* to see or hear.

Seth uses our experience of color as an example to highlight the complex mechanisms of perception at play. Our color perception seems to be so unimpeachably real (a red apple, a blue sky), but as we have known since Issac Newton's discoveries in the field of optics, this is not a 'pure,' direct reflection of what is actually there but a construction of the brain. In this sense it is not that seeing is believing, but that believing is seeing, and as a familiar maxim goes, "We see the world not as it is, but as we are."[11] Our experiences of the world around us are fully multi-sensorial and immersive, with our brains actively constructing our worlds to create our internalized cinematic experiences. Our brains (as well as similarly structured organs found in other creatures) are incessantly active—forever trying to predict the streams of incoming sensory stimulation before that sensory information even arrives, and actively constructing hypotheses to explain our experiences and fill in missing data. professor of philosophy Rick Grush elaborated on such findings, noting how this shift in emphasis suggests that perception is shown to be

> not a matter of starting with materials provided in sensation and filling in blanks until a completed percept is available. Rather, completed percepts of the environment are the starting point, in that the emulator always has a potentially

---

[10] Seth, "The Neuroscience of Reality."
[11] This maxim has unclear origin—it has been attributed to the Babylonian Talmud and Immanuel Kant. It also appears in a work by Anaïs Nin. See Nin, *Seduction of the Minotaur*.

self-contained environment emulator estimate up and running... The role played by sensation is to constrain the configuration and evolution of this representation. In motto form, perception is a controlled hallucination process.[12]

What this means is that, in a way, human brains are *always* generating hallucinations, which, although influenced by reality, are still hallucinations in the sense that they are not objectively true perceptions of reality. Clark, Hohwy, Seth, and Frith are all unified in their reference to this idea of perception as a process of active construction—which has come to be known by the phrase 'perception as controlled hallucination.'

Imagine being a brain. You're locked inside a bony skull, trying to figure what's out there in the world. There's no lights inside the skull. There's no sound either. All you've got to go on is streams of electrical impulses which are only indirectly related to things in the world, whatever they may be. So perception—figuring out what's there—has to be a process of informed guesswork in which the brain combines these sensory signals with its prior expectations or beliefs about the way the world is to form its best guess of what caused those signals. The brain doesn't hear sound or see light. What we perceive is its best guess of what's out there in the world.[13]

Seth, along with colleague Chris Frith, explains that our everyday perceptions can be understood from this perspective as being fantasies that sometimes—but not always—coincide with reality. And whilst we know that the hallucination idea is an old one, its implications are only recently becoming clear. We believe that we directly see, hear, and feel an objectively true external world. However, what these stories tell us is that the sensory signals delivered to our brains do not, on their own, fully reveal an external world but instead provide the raw materials used to generate the complex perceptual scenes that then guide our behavior. We now recognize that our experiences of the world around us are only indirectly related to an external reality. As Frith suggests, "our brains build models of the world and continuously modify these models on the basis of the signals that reach our senses", while noting that our perceptions may not be the world itself, yet they may as well be. Perceptions for Frith are "fantasies that coincide with reality,"[14] a realization that invites experimenters and artists of all stripes to exploit this fundamental ambiguity. We will examine many examples of this process through our exploration in this book of specific perceptual phantoms. Such phantoms can clearly reveal how our brains operate as predictive engines—as probabilistic prediction machines—forever guessing at the structure of the incoming sensory signals. So, if suggestions like those above indicate that perception is a *kind*

---

[12] Grush, "The Emulation Theory of Representation." Grush appears to be the first person to have mentioned this phrase in writing, but acknowledges that he owes the phrase to Ramesh Jain, who produced it during a talk at UCSD. We will also note further on that this idea can be traced to Hippolyte Taine in 1872.

[13] Seth, "Anil Seth."

[14] Frith, *How the Brain Creates Our Mental World*, 135.

of controlled hallucination, does this actually mean that perception *is* a controlled hallucination, full stop? Clark has often suggested that, indeed, it would be more important to think of hallucination as a kind of *uncontrolled* perception.[15] This is an important distinction, which will be relevant for us as we continue with our own story.

The more research that has been done into human perception as a whole, the more a portrait has emerged in which predictive brains appear to have been a fundamental fact of human evolution, minimizing the potential for danger and maximizing more favorable circumstances. We perceive our surroundings—and ourselves within them—not as they are, but in the modified or edited form that is most useful to us personally. Relating to such adaptive success (and also lending some credence to a concept of human brains as favoring energy conservation), neuroscientists and philosophers use terms such as 'neural frugality' or 'thriftiness' when describing our need to reduce the complexities of 'expensive' and complex neural processing. Biologically, we have limited resources and, as our brains require quite a significant chunk of those resources, it makes evolutionary sense that we should come to be as efficient and frugal as possible.

This, on its own, may not be enough to remove the aura of otherworldly, occasionally frightening power from the term 'hallucination' when and where we encounter it. Being familiar with various striking visual and auditory illusions, we mostly think of them as being the strange perceptual exception to the norm, but if these predictive processing accounts are essentially 'on track' (as Clark likes to put it) then we are in a situation whereby this general strategy is no exception, but rather one that actually informs all of human perception. Adding further insight, Clark contends that illusions could be best understood as being 'optimal percepts,' in the sense that they will be the best possible interpretation of the incoming signal under known circumstances (given what our systems already know about the world) and inevitably "a few local failures, then, are just the price we pay for being able to get things right, most of the time, in a world cloaked by ambiguity and noise."[16]

These facts of human perception have informed and inspired the creation of 'sonic phantoms,' a term that I coined and have been using to refer as a whole to a cohesive collection of sound compositions and related ongoing research first developed in 2009 and continuing into the present. Their creation has called upon a wide array of creative tools and means—musical instruments, voice, objects, and natural sounds—and has dealt at a fundamental level with perceptual auditory illusions of the type that would be projected by a 'predictive brain' framework as hinted at above. However, in terms of its potential application to compositional work other than mine, and, more specifically, in terms of the identification of the phenomenon, I introduce this term to describe a particular category of auditory illusions. Sonic phantoms not only 'trick' the auditory system by producing deceiving impressions of spatial or temporal experience, as other acoustic illusions do; they are auditory illusions that specifically give rise to the generation of illusory entities or patterns bearing the character of presences with apparent semantic or meaningful features (and as we progress through this volume, it

---

[15] Clark, *Surfing Uncertainty*, 308.
[16] Ibid., 51.

should become clear that these 'semantic' features are not exclusively 'linguistic' ones, a misunderstanding fairly common in an age where the 'linguistic turn' still exerts a formidable presence in philosophy and other intellectual disciplines). In other words, they manifest as a 'phantasmatic' *emergent* phenomenon playing with our innate human pattern-seeking propensities.

Sonic phantoms are apparently heard, but have no physical reality in the sense of being acoustic waves that act as perceptive inputs. They are auditory illusions that we can indeed perceive, but their locus of existence as audible phenomena is confined to our minds (or our brains, speaking in more strictly material terms). This occurs, in these particular cases, as a divergence between the acoustic properties of sounds and our subjective, 'phantom-like' perception of them. When we experience these illusions, we are hearing emergent 'fictional' sounds, which arise as a subjectively experienced result of a very particular organization of the 'real' acoustic elements present. Our story then focuses on creative exploits of playing with such sonic phantoms and on the various kinds of composed 'sonic elemental organizations' that lead to the experience of phantasmatic presence. To be sure, these sonic phantoms are not exclusively the result of a specialized compositional or acoustically trained ear (and, as we will show in more detail below, they are a natural [biological-cognitive] and universal [not culturally specific] phenomenon). Similarly, they are naturally one type among other possible, non-sonic, varieties of perceptual phantoms.

With different degrees of intensity and persistence, we all experience a perceptual phantom state of 'controlled hallucination' everywhere in the world around us. The 'inferred fantasy' view of perception insists that our conscious experiences arise from generative models that are nearly identical to virtual reality models. Unlike true fantasies or purely virtual constructs, conspicuous for their lack of conscious filtering of perceptive elements, this perceptual model is more of a compromise in which sensory input is being 'held at bay': and again, this is an evolutionary process carried out for functional, adaptive purposes. Non-semantic, 'noisy' percepts are minimized to the best of our abilities, though (as we will see soon enough) there is a fairly constant tug of war being played between prior beliefs and fresh sensory information—and, to be sure, this is not a process whose reliability ever perfectly stabilizes. Our biological systems' goal is for an overall optimized performance,[17] which explains why we might experience particular illusions, whether intentional or not. Getting it 'wrong' is the price we pay for mostly getting it right.

For obvious survival reasons, and in order to meet the optimization performance goals related to that survival, the objective of our everyday behavior is precisely to avoid the constant presence of the phantoms or illusions that can arise, and to develop highly structured systems or hierarchies of information that feature a minimum of confusing overlap or vagueness. In general, we are remarkably successful at doing so, and for good reason: there are definite evolutionary rewards associated with—to borrow the philosopher Casey O'Callaghan's rallying cry—'taking sounds seriously.' For example, in his treatise on 'sonic realism,' he hints at the exact (and regularly unappreciated)

---

[17] Lupyan, "Cognitive Penetrability of Perception in the Age of Prediction."

value of sounds for our general orientation and proprioceptive ability, claiming that sound "surpasses vision in the ability to detect change and in the ability to monitor multiple sources of information."[18] Yet, for all of the auditory system's sophistication in this regard, it should not be forgotten that we are aesthetically oriented creatures as well. Seen in this light, the auditory system is much more than a reliable tool for ensuring continued survival; it is just as useful in the act of shaping new subjective realities as it is in reporting on shared reality.

Neurological researchers Ramachandran and Hirstein remind us that "the purpose of art ... is not merely to depict reality—for that can be accomplished very easily with a camera—but to enhance, transcend, or indeed even to *distort* reality."[19] Phrases such as Coleridge's 'willing suspension of disbelief' suggest the degree to which this process is voluntarily entered into, and betray the fact that self-generated illusions are often essential to an aesthetic life that would be impoverished if it were dependent purely upon representational affirmations of existing reality. So it makes sense that, in sharp contrast to everyday life, our engagements with the different creative realms—for example, art, film, music—often involve deliberately creating various illusions of this nature, to 'trick' or fool the brain into seeing or hearing things that are actually not real and present. There is a considerable number of explanations for why we may choose to do this, ranging from staving off blank boredom, to enhancing our proprioceptive abilities, to much more sublime (and not necessarily conscious) attempts at communing and integrating with forces greater than ourselves. Likewise, there are manifold ways of engaging our auditory apparatus to do so, and many that require only minor adjustments to the listening process or to the production of sounds as we habitually experience them. Take, for example, the impressive number of effects arising from sounds' timbral features alone, and the way in which recordings of a trumpet note and a note played on an open violin string can become impossible to distinguish if one does nothing more technically complex than removing the 'onset' or 'attack' phases from recordings of these sounds. This exact procedure has been cited as the means with which to build up a storehouse of 'pseudo-instruments,' which has been one of the foundation stones of the *musique concrète* genre's development;[20] part of its program of "not responding to change, but suggesting it—provoking it."[21] A simple editing process such as this is often all that is needed to disrupt our customary association of certain sounds with unique timbral profiles and the speed at which they travel.

Entire expressive genres and cultural movements have followed from this realization that tricks played upon the listener might become the basis for enlightenment about humans' perceptual prejudices or biases, or might otherwise provide a sort of generalized inspiration by way of disorientation. Again, the strategies of much *musique concrète* and/or acousmatic music (which have their genesis in Pierre Schaeffer's work

---

[18] O'Callaghan, *Sounds*.
[19] Ramachandran and Hirstein, "The Science of Art."
[20] "An accidental finding at the early stages of research into noises contained all the seeds of *musique concrète* (Schaeffer 1952, p. 16): a bell sound was recorded after its attack, being thus rendered unrecognizable." Palombini, "Machine Songs V."
[21] Emmerson, "From Dance! To 'Dance.'"

at the sound effects warehouse for French radio) have a strong reliance upon instilling a sense of uncertainty about what one is hearing, and of making implausible, fantastic hybrids of natural and synthetic worlds seem convincingly 'real.' Musical instrument and software designers and laboratories, likewise, have put considerable work into innovations such as the phase vocoder (a digital sound analysis/synthesis tool whose 'cross synthesis' feature enables something very much like that hybridization just now described), ensuring that consumer-level technology will steadily increase the degree to which we can daily generate or interface with new sonic illusions.

Indeed, as this narrative unfolds, we will find time and again that music distinguishes itself as the practice of creative work with sound most likely to give rise to the richest and more open forms—unguided, non-descriptive, non-literal—of 'auditory fictions' being the fertile soil from which the phantasmatic grows. Numerous philosophers, from Arthur Schopenhauer to Roger Scruton, have identified music as being an art form that is completely expressive of subjective human experience rather than being simply representational, and have consequently argued against music being a kind of experience separate from itself (this was indeed the main tenet of the Romanticists' aesthetic program of 'Absolute Music'—remarkable in its historical exceptionality— from the late eighteenth century until perhaps the early twentieth century).[22] One of the results of this philosophy reaching a larger consensus has been that many of the other art forms find themselves "aspiring to the condition of music"—as essayist Walter Pater suggested in 1877.[23] This is a realization that has far-reaching conclusions, which at the very least makes the understanding of phantasmatic sound meaningful for creators who work outside of the sonic and musical fields.

A multitude of cultures worldwide and throughout the ages have created music with very different techniques that nonetheless share a common feature of exploiting our pattern-seeking tendencies in order to intentionally generate striking sonic phantoms. These kinds of phenomena, as well as their creative exploitation, have indeed been recognized, described, and researched independently in different fields of study and musical practice. Ethnomusicologists such as Gerhard Kubik have described particular forms of instrument-generated sonic phantoms as "inherent or subjective patterns" of hearing.[24] Cognitive psychologists and psychoacousticians such as Albert Bregman have described and researched them in terms of 'auditory streams.'[25] And within the realm of music composition, terms such as 'emergent patterns,' 'resultant rhythms,' or 'pseudo-polyphony' are common when referring to different varieties of them.[26]

It has been widely researched, both theoretically and practically, that expectation and prediction make up some of the central mechanisms in the perception and cognition of music. It has been suggested that music perception "is an active act of listening, providing an irresistible epistemic offering... when listening to music

---

[22] Dahlhaus, *The Idea of Absolute Music*.
[23] Pater, *The Renaissance*.
[24] Kubik, "The Phenomenon of Inherent Rhythms in East and Central African Instrumental Music."
[25] Bregman, *Auditory Scene Analysis*. See also McAdams and Bregman, "Hearing Musical Streams."
[26] Jones, *Studies in African Music*; Reich and Hillier, *Writings on Music, 1965–2000*; Pandey, *Encyclopaedic Dictionary of Music*.

we constantly generate plausible hypotheses about what could happen next, while actively attending to music resolves the ensuing uncertainty."[27] Music that is largely based on predictable regularities of a temporal, harmonic, or textural nature presents an intriguing art form for researchers to explore and to understand the fundaments of the predictive brain, being particularly "a powerful tool to investigate PC in the brain, because the statistical regularities in music are so well defined... thus we have seen that the PC framework can account for several key phenomena in auditory processing."[28] Such research using musical structures can throw a spotlight on how our underlying perceptual mechanisms make use of feature-based and temporal prediction models.

In light of the fact that our general mode as predictive organisms is to avoid 'prediction errors,' it is often the case that music paradoxically pulls us in the opposite direction, favoring or encouraging the violation of expectations. However, we know that when we listen to music we happily entertain all kinds of hypotheses about how a musical work will unfold at all kinds of time scales. Music can provide innumerable opportunities to resolve sonic uncertainties at various levels, and so tensions and their resolutions (or lack thereof) can be enjoyed and their listeners rewarded.

1.2 Presence by Apophenia

HAMLET
Do you see yonder cloud that's almost in shape of a camel?
POLONIUS
By th' mass and 'tis, like a camel indeed.
HAMLET
Methinks it is like a weasel.
POLONIUS
It is back'd like a weasel.
HAMLET
Or like a whale.
POLONIUS
Very like a whale.

—*Hamlet*, Act 3 Scene 2

... *nothing is so alien to the human mind as the idea of randomness.*

—John Cohen

---

[27] Koelsch, Vuust, and Friston, "Predictive Processes and the Peculiar Case of Music."
[28] Ibid., 74 (PC here is Predictive Coding).

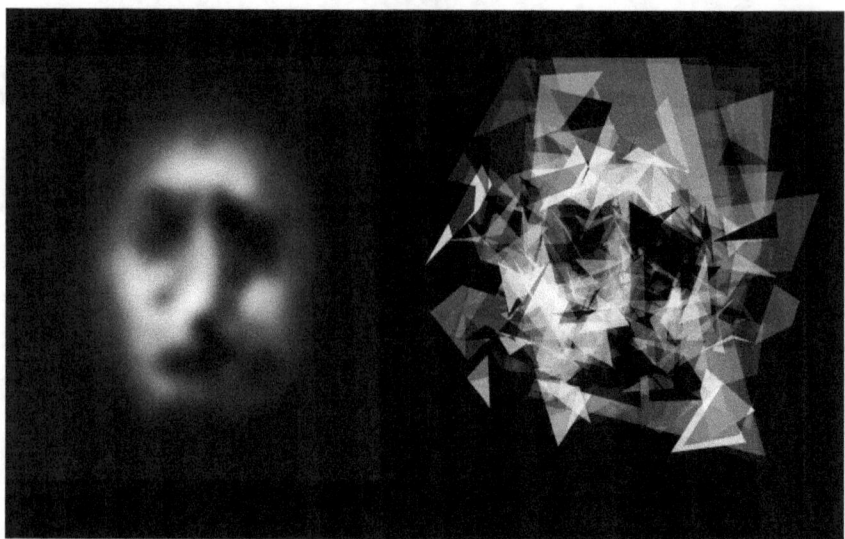

Figure 1.2 *Phantom Portrait*. Human perception is dominated by pattern recognition, which naturally and regularly induces pareidolia, or the psychological phenomenon of perceiving patterns in randomness. Computer systems and AI are also inevitably imbued with pareidolia when given cognitive capacities such as facial recognition. In my visual project *Phantom Portraits* (not included in this book) I artistically explore this techno-perceptive situation by employing face-detection algorithms to analyze, select, and evolve accretions of randomly generated polygons (via the Pareidoloop open-source software) until the computer 'sees a face.' This figure shows a prototypical example of such a 'face' after some 'polishing' with just a center edit and slight blurring (left) of an evolved polygonal 'face' (right). Image by Barbara Ellison, 2014.

The neurological phenomenon of misfiring or misinterpreting patterns, that is to say, making connections and attributing abnormal significance to them from meaningless random data or formless assemblages of sensory inputs, was given the name of 'apophenia' by the German neurologist Klaus Conrad,[29] with the term 'pareidolia' referring more specifically to the tendency to apply this to imagery or sounds. The esotericist Peter Carroll provides a perhaps more elegant comparison of the two concepts as follows:

> Apophenia means finding pattern or meaning where others don't. Feelings of revelation and ecstasis usually accompany it. It has some negative associations in psychological terminology when it implies finding meaning where none exists, and some positive ones when it implies finding something important, useful, or beautiful. It thus links creativity and psychosis, genius and madness ... It has close

---

[29] The term is perhaps a misnomer incorrectly attributed to Klaus Conrad by Peter Brugger, who defined it as the "unmotivated seeing of connections" accompanied by a "specific experience of an abnormal meaningfulness," but it has come to represent the human tendency to seek patterns in random information in general (such as with gambling, paranormal phenomena, and religion). Conrad, *Die beginnende schizophrenie*; Houran and Lange, *Hauntings and Poltergeists*.

associations with pareidolia, the mistaking of pieces of rope for snakes, seeing goats, bulls and virgins in the positions of stars and in the personalities of people ...[30]

In our view, as well, apophenia with a clear link to the predictive brain is likely one of the prime mechanisms for the generation of the kind of phantasmatic 'presence' that we referred to above. Apophenia derives from the Greek 'Apo,' which means 'away from,' and 'Phenia/Phren,' which refers to the mind or cognitive faculties. Examples of it abound: we see faces and meaningful forms in clouds, rocks, or foodstuffs; and in such cases we cannot help the unconscious impulse to endow those meaningless random forms with significant meaning. Our own graphic examples (Figures 1.1 and 1.3) come from a very humorous Irish publication featuring a selection of snaps (submitted by *Broadsheet* readers) of pareidolic maps of Ireland, entitled *The Broadsheet Book of Unspecified Things That Look Like Ireland*. On the book's cover is written "There is only one litmus test for Irishness, and that is the eagle-eyed ability to detect the island's shape wherever it occurs in nature ..."[31]

Examples such as our pareidolic maps of Ireland are typically viewed with a sense of whimsy, or perceived as quaint artifacts that do not make any deeper commentary upon our psychological inclinations. However, there is a decidedly more serious side to this, as personified by Rudolf Arnheim, one of the most prominent individuals to apply Gestalt[32] psychology to the arts. Arnheim was no stranger to the prevalence of illusionism in those arts, and notably claimed in his signature work *Art and Visual Perception* that even the most apparently simple, non-communicative marks or etchings on a surface can transmit meaning to the viewer.[33] From this unequivocal statement came a host of associated realizations, not the least of these being that the power of visual suggestion can turn even flat lines into "one-dimensional objects ... as though ... wrought in iron or some other solid material."[34] It would seem that a determined striving towards meaning, or even a 'narrative' quality, is something that we do in fact attribute to the whole phenomenological world: moreover, we seem to derive considerable pleasure or satisfaction from being able to make "*objects* appear more vividly as *things*," as theorist Jane Bennett suggests. This drive towards semantization is no less true of audio information, as is maybe exemplified by the absurd misheard lyrics (known as 'mondegreens'),[35] and humorous translations of undecipherable

---

[30] Carroll, *Apophenion*, 8.
[31] Coughlan, *The Broadsheet Book of Unspecified Things That Look like Ireland*.
[32] Rudolf Arnheim (1904–2007) was a German-born author, art and film theorist, and perceptual psychologist. He learned Gestalt psychology from studying under Max Wertheimer and Wolfgang Köhler at the University of Berlin and applied it to art.
[33] Arnheim, *Art and Visual Perception*.
[34] Ibid., 219.
[35] Mondegreens are a kind of aural malapropism where one mishears a lyric. The word 'mondegreen' was coined inadvertently by the American writer Sylvia Wright after mishearing the words of the Scottish ballad *The Bonny Earl of Moray*. She wrote an essay entitled "The Death of Lady Mondegreen," published in *Harper's Magazine* in November 1954 (and featured in our epigraph). The misheard lyric was originally "And Layd him on the green." Wright, "The Death of Lady Mondegreen."

foreign language songs into the hearer's native tongue (popularly known by the Japanese term *soramimi* [loosely 'ears in the sky']); though perhaps knowing full well from a song's melodic cues that it deals with some vague yearning or unrequited love, it is still irresistible to warp the phonetic content into our primary language and in the process make that song's phonetic content into, say, instructions for food preparation with emotionally incongruous instrumental backing.

While such a perceptive quirk might be a commonly reported example of semantization among adults, we should also consider the theory that these tendencies are more or less always with us, forming extremely early in human lives and thus deepening the emotional resonance that comes with discovery of familiar information secreted within unfamiliar percepts. Studies on infants' responses to faces, as well as their reactions to objects with face-like components, have strongly suggested that this perceptive ability is 'hard-wired' within us—a point that has historically been brought to the fore by scientific skeptics such as Carl Sagan in their arguments against mystical thinking.[36] The Doshisha University researchers Masaharu Kato and Ryoko Mugitani found that "in brain activity measurements, the same event-related brainwave component was observed during the presentation of images of real faces and pareidolic faces."[37] When they presented infants with an abstract, top-heavy image in which four black blobs were set, infants stimulated by a sound were inclined to look at the bottom-most 'blob,' that is, that which most would visually associate with a 'mouth' (the middle blobs were spaced apart in such a way as to represent 'eyes,' while the top-most blob was intended as a 'distractor' not mappable to human facial features). Curiously, this occurred when a pure electronic tone of a 400 Hz frequency was used in the experiment, rather than an actual human voice: this is a fact that seems to strengthen the argument of infants' pareidolic perception rather than weaken it, given that Kato and Mugitani's assessment that "infants 'knew' in advance about the existence of the mouth through the configuration of blobs and contour, and thus looked at the bottom blob as the most relevant area for this sound."[38]

With perception now understood to be an active constructive process, we know as cognitive creatures that successfully interpreting our environment relies upon actively searching for patterns that are meaningful to us in terms of structure and information. The brain is, at every moment, actively processing sensory input as we try to make meaningful sense of what is going on in our immediate environment, and in the version of reality we comprehend by means of technological extensions. We 'see as …' or 'hear as …' because the world around us is—as philosopher William James famously put it—a jumble of "blooming, buzzing confusion." For many, we experience the world around us 'through a glass, darkly,' as in the blurry reflections of brass mirrors of antiquity.

[36] Per Sagan, "As soon as the infant can see, it recognizes faces, and we know that this skill is hardwired in our brains … the pattern-recognition machinery in our brains is so efficient from extracting a face from a clutter of other detail that we sometimes see faces where there are none. We assemble disconnected patches and dark and light and consciously *try* to see a face." Sagan, *The Demon-Haunted World*, 45.

[37] Kato and Mugitani, "Pareidolia in Infants," 1–9.

[38] Ibid.

Figure 1.3 *The Dead Pigeon on a Branch That Looks Like Ireland* (top image, photograph by Tadhg O'Halloran) and *The Brillopad That Looks Like Ireland* (bottom image, photograph by Will Hanafin). Kind permission by Broadsheet.ie, publishers of the online and printed source image in the book *The Broadsheet Book of Unspecified Things That Look Like Ireland*, edited by Aidan Coughlan, 2013.

If our view of memory is correct, in higher organisms every act of perception is, to some degree, an act of creation, and every act of memory is, to some degree, an act of imagination.[39]

The neurological basis for such an orientation was also memorably stated by cyberneticist W. Ross Ashby, who, in claiming that "the whole function of the brain is summed up [as] error correction" may have been unnecessarily bold but not entirely false.[40] This is to say that our means of interpreting the world are radically different from the simple 'feed-forward' process in which, for example, sensory data strikes the retina and transmits directly to the brain, being fed through multiple levels of cortical processing and ultimately triggering an appropriate motor reaction with no additional preliminary or modulating processes. If instead we proceed from the predictive processing or 'Bayesian' model of perception (after mathematician/theologian Thomas Bayes), then our Bayesian brains process data in a less simplistic manner than this. That is to say, our brains predict what our eyes will see even *before* we process any sensory data from the retina. Clearly, we are dealing with a perceptual experience more complex than the 'feed-forward' model, since there is the additional step of basing our assimilation of new data upon previous experience. Similar to the formulae that define other types of Bayesian networks, the 'Bayesian brain' operates on a principle of "initial belief + new evidence = updated belief."[41]

This model of perception accounts for the complete spectrum of human behavior, from simple interactions with physical reality to the formation of more complex cultural beliefs (i.e., the valuation of 'authenticity' stemming from the ease with which the behavior of 'authentic' personalities can be predicted). As far as an unimpeachable scientific consensus goes, the jury is still out on this, but if our brains really turn out to be Bayesian, meaning that predictive processing is *the* fundamental principle of cognition, then ultimately it means that all our perceptive and cognitive activity is a matter of making predictions. If it is not already evident how comprehending reality this way can influence imagination and subsequent creation, this all neatly dovetails into a more art-historical context via the concept of 'the beholder's involvement,' a term introduced by the Austrian art historian Alois Riegl[42] and later elaborated and

---

[39] Edelman and Tononi, *A Universe of Consciousness*, 101. This quote is also attributed to Oliver Sachs.
[40] Clark makes the note that this remark is simply described as a "scribbled, undated, aphorism" in the online digital archive of the scientist's journal; see http://www.rossashby.info/index.html. In Clarke, "Whatever Next?", 181.
[41] Although this is a complex subject, for our purposes here, the Bayesian brain, predictive processing, hierarchical predictive coding, and other titles are more or less describing the same story whereby experience is constructed from an ever-changing balance between sensory evidence and expectations (or top-down predictions). For a nice primer, see Wiese, Metzinger, and Group, "Vanilla PP for Philosophers a Primer on Predictive Processing."
[42] Alois Riegl (1858–1905), a member of the Vienna School of Art History, was one of the major figures in the establishment of art history as a self-sufficient academic discipline, and one of the most influential practitioners of formalism.

popularized by Ernst Gombrich as 'the beholder's share'.[43] Riegl's insight was that art is incomplete without the perceptual and emotional involvement of the viewer and that part of the meaning of the artwork is contributed by the viewer. Gombrich, meanwhile, essentially posits that perception can be understood as a *generative* set of processes owing to the ground rules laid out above: once we take this as a given, and then realize that we are generating ambiguous and imperfect versions of the world whether acting as cultural transmitters or cultural receivers (e.g., an audience), a form of creativity results wherein these two ambiguities interface and the interpretation of the original creative act becomes a creative act in itself. Gombrich's Viennese colleague in art criticism, Ernst Kris, was also a key proponent of this idea, and his thoughts conveyed by Eric Kandel on the subject are worth reiterating here:

> Kris argued that when an artist produces a powerful image out of his or her life experiences and conflicts, that image is inherently ambiguous. The ambiguity in the image elicits both a conscious and unconscious process of recognition in the viewer, who responds emotionally and emphatically to the image in terms of his or her own life experiences and struggles. Thus, just as the artist creates a work of art, so the viewer re-creates it by responding to its inherent ambiguity. The extent of the beholder's contribution depends on the degree of ambiguity in the work of art… Kris argues that ambiguity enables the artist to transmit his own sense of conflict and complexity to the viewer's brain.[44]

For many, this shift in aesthetic and perceptual attitude will be something entirely new, yet for Kris and Gombrich, and their contemporary acolytes like Anil Seth, this is the way that things have always been: art is inconceivable without the 'beholder's share.' The latter has pointed at specific genres that provide confirmation of this theory, such as Impressionist painting and its feted means of rendering objects in the periphery of vision (in particular, Camille Pissarro's work is lauded for its "'reverse engineering' the visual system" and "[recovering] the afferent sensory signals that trigger a particular cascade of sensory inference, rather than depicting the outcome of this process").[45] Such achievements could be viewed as an elegant proof of the Bayesian/predictive brain model, and they are far from being unique in that respect: the sonic realm is equally capable of displaying this to us while simultaneously allowing us to make fresh use of it.

Keeping in mind the imperfect state of our total perception hinted at above, we are often confronted by a number of possible and plausible interpretations for our sensations. The more ambiguous the scene, the more actively we search to find patterns, and therefore a greater number of possible and plausible patterns are there to be detected. This manifests in a particularly vivid way with what are known as bistable or multistable illusions, the most commonly recognized of which are ambiguous

---

[43] Ernst H. Gombrich (1909–2001) was an Austrian-born art historian who was one of the field's greatest popularizers, introducing art to a wide audience through his best-known book, *The Story of Art* (1950).
[44] Kandel, *The Age of Insight*.
[45] Seth, "From Unconscious Inference to the Beholder's Share."

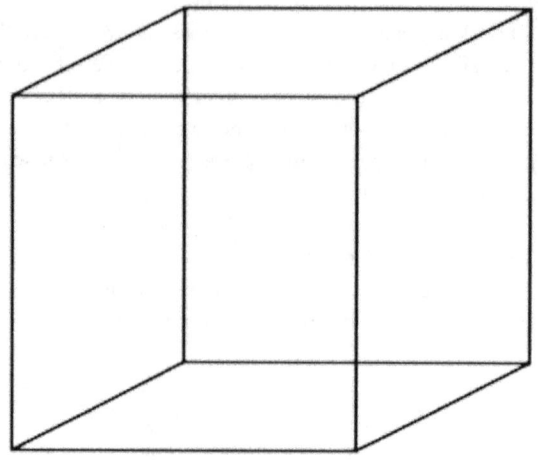

Figure 1.4 The 'Necker cube' and the 'Rabbit and Duck,' classic examples of perceptual bistable images. The Necker cube is an optical illusion first published as a rhomboid in 1832 by Swiss crystallographer Louis Albert Necker. The simplicity of this wire-frame drawing of a cube gives no visual cues as to its orientation, so it can be alternatively interpreted to have either the lower-left or the upper-right square as its front side. The classic 'Rabbit and Duck' (*Kaninchen und Ente*), first published by psychologist Joseph Jastrow in 1899, is also a reversible image that induces a constant perceptual back-and-forth between the duck and the rabbit.

figures such as the Necker cube and Jastrow's famous 'Rabbit and Duck' image, which, true to its name, evokes two very different and equally valid visual experiences: that of a duck or that of a rabbit.

Multistable illusions such as in Figures 1.4, 1.5, and 1.7 play with our everyday sense of perception in that they can elicit multiple perceptually valid interpretations. In this way, they give lie to the previously discussed idea of the 'world as controlled hallucination,' revealing how easily our own brains can fool us and reminding us that we cannot always rely on our senses for an objective and accurate representation of the world. If we are to take recent findings in neuroscience seriously, then there is no such thing as an objectively accurate representation of the phenomenal world. If the reality of this chaos and unreliability causes trepidation in the reader, it might be comforting to also consider this: despite the seeming universality of the neural processes that bring about this 'controlled hallucination,' each individual will have his or her own version of the 'controlled hallucination' inner reality model which is wholly personalized and unique.

As the saying 'natura abhorrent vacuum' ('nature abhors a vacuum') implies, even in the case where there is 'nothing,' we will endeavor to find 'something,' emphasizing that our brains are predisposed towards seeing 'something' rather than 'nothing.' Our brains function as pattern-detection and prediction machines, connecting dots to find meaningful relationships within the constant bombardment of multiple sensory inputs that we have to deal with. Naturally, without this remarkable ability—'patternicity,' as described by Michael Shermer—the world around us would be utterly chaotic and overwhelming.

Numerous anthropologists and researchers on the origins of religion and art see our pattern-seeking propensity—a remarkably magnified ability considered to be intrinsic to perception in humans—as a preeminent natural basis of anthropomorphism, animism, religious beliefs, superstition, and magical thinking. Professor David Lewis-Williams, a cognitive archeologist, writer, and expert in Paleolithic rock art and Neolithic monuments, has developed a neuropsychological model, describing a neurology of mystical experience—directly related to the notion of apophenia described earlier—that theorizes that humans have the propensity for illusory aural and visual experiences, which we typically feel as mystical experiences, because of the way our brains are 'wired': "The ghost hidden in the machine is a cognitive illusion created by the electro-chemical functioning of the brain."[46]

Particularly as it relates to the community of scientific skeptics and would-be banishers of paranormal explanations for everything from UFO to prophecy, the significance of apophenia lies in its ability to clearly display our universal and limitless capacity for self-deception. Moreover, within the skeptics' community, the understanding of apophenia as a form of "*post hoc* data analysis … spotting clusters of results and giving them an undue significance"[47] is often used as a prelude to a more sweeping program of unveiling 'wishful thinking' and 'magical thinking,' and the sort of confirmation bias exploited

---

[46] Lewis-Williams, *The Mind in the Cave*, 105.
[47] McNulty and Williams, "If You're Going to Vegas," 21.

by marketers and cults alike. The 'Texas sharpshooter' fallacy, in which a hypothetical gunman fires a number of aimless shots at the side of a barn and then draws a typical target pattern of circles around the 'tightest group' of shots, is one such celebrated example of apophenia as deliberate self-deception and, in turn, attempted deception of the public. Dozens of other cautionary examples have been mobilized by skeptics as a general warning against trusting our subjective experience as the final judgment on any issue, and they have been used more specifically as a precaution that "individuals can stress similarities in data rather than differences,"[48] selectively focusing only on the most meaningful clusters of information while ignoring the vast and often contradictory informational landscape that surrounds them. From a purely biological point of view, apophenia in general would be a 'better-safe-than-sorry' strategy: in terms of survival, it is safer for us to wrongly attribute organization, structure, or 'liveness' to inanimate things than not to detect a life-threatening predator. As Guthrie put it: "it usually is less costly to mistake boulder for bear than to mistake bear for boulder."[49]

With this in mind, it is worth further examining Lewis-Williams's proposal[50] that the realm of the supernatural is an illusion actually created by our neurochemistry, that is, the brain itself. His neuropsychological model describes how, through a process of natural selection, we have evolved our propensity for experiencing states of consciousness (mystical, trancelike, supernatural) as a result of our 'neurological hard-wired foundation.' All our 'normal' and intensified mental states would be generated by the neurology of the nervous system, and would therefore be a fundamental part of being human in the biological sense. Cultural contexts may of course diminish or amplify their effect, but the biological basis would always be there.

Before going much further, these 'cultural contexts' deserve some additional examination. Despite its universal nature, the actual content of any apophenic 'phantom' presences will be naturally modulated all over the world due to the diverse social-cultural environments. This is something that should be expected if one looks into the culturally distinct variations on 'phantom' presences that occur in other liminal states of consciousness: the condition of hypnagogia (i.e., the often hallucinatory state occurring between complete wakefulness and sleep) provides just one such example. Researcher Andreas Mavromatis, to whom the condition's name is attributed, notes certain pareidolic universals within the reports of those experiencing it: he notes that "the seeing of faces is so universal among the hypnagogic imagers that, as Leaning put it, 'it almost suggests that there is a certain "face-seeing" propensity within the mind,'"[51] and goes on to note how these faces appear not fully-formed at first sight but evanesce into being ("faces appear in a variety of ways but most often they seem to take shape out of a misty stuff, sometimes one face forming from another").[52] Another example of this tendency is provided by Richard Gregory's well-known "hollow mask" illusion,

---

[48] McNulty and Williams, "If You're Going to Vegas," 21.
[49] Guthrie, *Faces in the Clouds*, 80.
[50] Lewis-Williams, *The Mind in the Cave*; Lewis-Williams and Pearce, *Inside the Neolithic Mind*.
[51] Mavromatis, *Hypnagogia*.
[52] Ibid.

Figure 1.5 *Suède paréidolie*. A great example of both pareidolia and bistability. This photo is of a Victorian couple in the early 1900s, with a small child sitting on the knee of the man. The pareidolic alternative bistable image is a bearded man's profile with curly hair (a bush in the photo's background). Image (public domain) submitted by Jessica Lundgren, Sweden (a photo she got "from her grandma") to https://paranormal.about.com

which is still widely used to demonstrate just how powerful our human propensity for seeing faces can be.[53] Given this face-seeing bias, any bit of information that suggests a face, actually present or not, can easily lead us to find one.

Yet, at the same time, there is content that seems not to transfer between cultures: the common episodes of hypnagogic terror, in which subjects feel a malicious presence to be hovering over or sitting on top of them, are typified by regionally distinct variants such as ghostly foxes and *kanashibari* in Japan or 'night hags' or incubi in Western cultures. This holds very much true for pareidolic perceptions encountered in the waking state: while the Inuit may have visions of polar bears and seals in their environment, Irish Catholics may find the Virgin Mary (or Jesus) in theirs.

All of these phenomena can be explained within the context of Predictive Processing and Bayesian brain frameworks: Frith, Hohwy et al. suggest that culture in a very wide sense could be seen as a tool for precision optimization through shared context and that ritual, convention, and shared practices enhance mutual predictability between people's hidden mental states.

> If alignment of mental states is an integral part of how culture optimizes precision and communication efficiency, then culture should be seen as providing a set of frameworks for interpretation, rather than merely for scaffolding interpretation. If the brain is a hierarchical Bayesian network providing a perceptual fantasy of the world, then culture determines and constrains the hyperpriors needed by such a neural system.[54]

1.3 Apophenia and Creativity

A mild case of apophenia is a novelist's secret weapon that brings readers and literary success. We spend our working days seeing spontaneous connections between unconnected events, people, and lives, and weaving meaning into those connections.
—Christopher G. Moore[55]

Homo sapiens are about pattern recognition, he says. Both a gift and a trap.
—Cayce Pollard[56]

The usefulness of apophenia and pareidolia extend beyond their ability to hold up a mirror to the negative or counterproductive qualities mentioned above: logically

---

[53] Gregory, *The Intelligent Eye*.
[54] In response to Clark, "Whatever Next?" Paton et al., "Skull-Bound Perception and Precision Optimization through Culture."
[55] Moore, "Apophenia."
[56] Gibson, *Pattern Recognition*, 22–23.

and perhaps inevitably, this outstanding heightened capacity, natural and profoundly human, can also be a paramount source of creativity. As we generate beliefs and superstitions, so we can develop works of art, literature, or music that stem from apophenia. It could be said that the apophenic pattern recognition processes involved in pareidolia form much of the basis for all the expressive forms we understand as art: devices such as the poetic use of metaphor derive from an ability to reveal the occluded similarities between two seemingly dissimilar entities (a process that can then be applied to 'synesthetic' or cross-modal experiments in transferring the key semantic components and compositional elements between creative media). Most of the time, we experience our brains' Bayesian best guesses (our controlled hallucinations) as being real, but it is clear that the commonality of our experiences is only so complete, and in fact our experiences of 'realness' should not be taken for granted.

We know that a great deal of disagreement about common perceptual worlds exists, as proved in recent history by viral internet sensations such as 'The Dress.'[57] Individuals

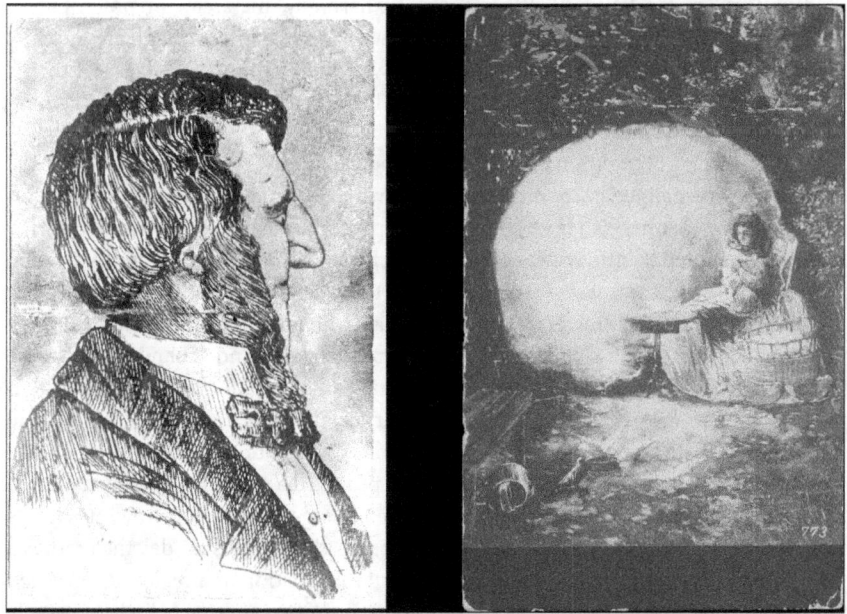

Figure 1.6 Bistable pareidolic images. Left: portrait of a man (Berlioz?) and naked woman (public domain; found at https://www.delcampe.net/fr/forum/cartes-postales/494018-identification). Right: *Metamorphic Postcard*, c. 1900. Wellcome Images, courtesy of the Richard Harris Collection, photo number L0071968 (this file is licensed under the Creative Commons Attribution 4.0 International).

[57] 'The Dress' was a photograph that became a viral internet sensation on February 26, 2015, when viewers disagreed over whether the dress pictured was colored blue and black, or white and gold. The phenomenon reveals differences in human color perception and was also much discussed within neuroscience circles. Wallisch, "Here's Why People Saw 'the Dress' Differently."

diagnosed with synesthesia have extra-sensory experiences such as vividly perceiving colors when hearing sounds or hearing sounds when seeing colors—experiences that are vivid, but are accompanied by a sense of unreality. It has been suggested that "synesthesia is 'abnormal' only in being statistically rare," and is indeed a normal brain process, but one only consciously available to some of us.[58] Such experiences of synthesis, illusions, apophenia, and so on reveal the creative constructive quality of perception.

Artist Jane D. Marsching defends those who nurture pareidolia in their daily lives and insists against the abnormality of such, claiming that we all "take spotty data and follow our desires through a chain of analogy to our end goal, along the way overlooking, shoving aside, or simply forgetting information and conclusions."[59] Elaborating on this, Marsching states the following:

> When I shut my eyes and look at the darkness, even nothing becomes something: a star field, a fractal simulation, an indistinct halo. The fact of the real is another world from the figment of our imaginations. Analogy groups the discordant and disordered into relationships forged not on harmony, hierarchy, or order, but instead on the chance connections of experience, imagination, and presence. Analogy satisfies our desire to place ourselves within the world—we link the known with the unknown to create an order that is dynamic and self-reflecting.[60]

The art-historical tradition that Marsching touches upon is also affirmed by Georges Sorel, who, writing in 1908, admonished those with an ideological bent to "look upon art as a reality that begets ideas ... not as an application of ideas,"[61] a wish that seems like it was granted by the arrival of abstract art a few years later. This propositional type of artwork, if it was not available in Sorel's time, certainly exists now, with thanks due in part to the near-universal condition of apophenia and to the aforementioned neurological phenomenon of the 'beholder's share' that goes hand in hand with it. There is an uncanny ease with which these phenomena can be bonded to a larger program of viewing artworks or listening to music as participatory and essentially 'unfinished' until the public or the audience joins in as secondary 'artists' or 'composers.' Again, Anil Seth's neurological research into perception has provided a most cogent explanation of this phenomenon of perceiver-as-artist, or of art as a generative act, which is closely related it to the 'top-down' model of cognitive processing. In this model, the brain is a kind of inference organ reliant upon background information or already accumulated knowledge (contrast this with the bottom-up method of processing, which essentially posits that sensation and perception act in concert, that is, there is no need for the additional interpretive step required by top-down processing).

---

[58] Cytowic, *Synesthesia*; Bailey, *To Hear the World with New Eyes*.
[59] Marsching, "Orbs, Blobs, and Glows."
[60] Ibid.
[61] Quoted in Read, *To Hell with Culture, and Other Essays on Art and Society*.

Seth is far from an isolated case in proposing such a perspective: elsewhere, neuroscientist Ernst Kandel in his book *The Age of Insight* memorably wrote "the brain is not a camera but a Homeric storyteller,"[62] with an understanding of the brain as an inference machine, generating hypotheses and fantasies that are tested against sensory data. Herein Kandel explicitly discusses 'the beholder's share' in direct relation to Helmholtz and his ideas of the brain being an organ of perceptual inference. Karl J. Friston,[63] in an article reviewing the book, perceptively writes,

> Put simply, the brain is—literally—a fantastic organ (fantastic: from Greek phantastikos, able to create mental images, from *phantazesthai*). For me, the story starts with Hermann von Helmholtz (1821–94) and the notion of unconscious inference (Helmholtz, 1866/1962). Kandel places this story in the context of history and art, in an illuminating and compelling way.[64]

Another highly relevant axiom, conveyed via Kandel and useful for the easy absorption of these concepts, comes courtesy of Gombrich who argued that,

> there is no 'innocent eye': that is, all visual perception is based on classifying concepts and interpreting visual information. One cannot perceive that which one cannot classify, ... He appreciated the role that *cognitive schemata*, or internal representations of the visual world in the brain, arguing that every painting owes more to other paintings the viewer has seen than it does to the world actually being portrayed.[65]

Put more succinctly, the viewer "recreates [a work of art] by responding to its inherent ambiguity," with the "extent of the beholder's contribution depend[ing] upon the degree of ambiguity in the work of art."[66] In doing so, his or her anticipation of certain aesthetic elements' presence tends to play a greater role in the perception of the artwork than what Gombrich identified as the power of conceptual knowledge. What is clear from such statements from Kandel, Gombrich, and Friston is that ambiguity or disambiguation is at the heart of perceptual inference: given that this is intrinsic to us, it is not surprising that it is mapped onto our interactions with technology.

Within the modern experimental media landscape, there is a rich body of work that, whether it does so willfully or not, speaks to the apophenic impulse in humans. Communication technologies have always carried with them an ability to be 'misused' as tools for active self-discovery rather than as instruments whose output is meant to be passively absorbed. Consider, for starters, that generally any form of audio

---

[62] Kandel, *The Age of Insight*, 351.
[63] Friston is responsible for the 'free energy principle' (active inference in the Bayesian brain). He has revolutionized and has had a major influence on human brain studies and has developed many innovating techniques in brain imaging. Karl, "A Free Energy Principle for Biological Systems."
[64] Friston et al., "Computational Psychiatry."
[65] Kandel, *The Age of Insight*, 204, 212.
[66] Ibid., 192.

amplification can act as a Very Low Frequency (VLF) receiver, and can therefore produce odd alien signals from within the VLF waveband, as any audio engineer will tell you (it is, after all, in their job description to minimize such phenomena from interfering with performances and broadcasts). As we will soon investigate, these intrusive signals have often taken on a life of their own once they have been seen by their listeners as 'communicative' rather than simply intrusive.

Pareidolic perception has proven remarkably adaptable to creative acts, which do not have to be any more elaborate or involved than the act of sculpting televised 'snow' on dead channels (or doing the same with audible radio static). Whatever the levels of complexity associated with apophenia-inspired activities, it is clear that there is an intellectual tradition of viewing apophenia and creativity as being intimately, organically connected. Neurologist Peter Brugger elaborates on this:

> The propensity to see connections between seemingly unrelated objects or ideas most closely links psychosis to creativity. Indeed, with respect to the detection of subjectively meaningful patterns, apophenia and creativity may even be conceived of as two sides of the same coin. One must keep in mind, however, that the term detection as used here does not refer to a process of mere identification, to finding the solution to a perceptual puzzle. Rather, the assumption of meaningfulness in randomness always involves a subjective interpretation of spatial and temporal configurations. The creative arts acknowledge and take advantage of this purely subjective act of perceiving. They have always been inspired by chance and randomness to create works of art whose meaningfulness is, however, left to the interpretation of its viewers.[67]

Taken to extremes, a voracious appetite for pattern detection draws dangerously close to states of paranoia and psychosis. Marked apophenia is often attributed to high levels of dopamine in the brain, which affect the propensity to find patterns and significance where there are none, and in acute cases is treated as a positive symptom of psychosis and schizophrenia. However, whilst in the extreme it can be indicative of a psychotic condition, on a more normal level it is now known to be a ubiquitous feature of everyday human experience, and for the open and creative mind it can indeed be a kind of 'happy genius':

> More conservative minds deprive coincidence of meaning by treating it as background noise or garbage, but the shape-shifting mind pesters the distinction between accident and essence and remakes this world out of whatever happens. At its obsessive extreme such attention is the beginning of paranoia (all coincidence makes "too much sense"), but in a more capacious mind it is a kind of happy genius, ready to make music out of other people's noise. Either way, the intelligence

---

[67] Brugger, "From Haunted Brain to Haunted Science."

that takes accidents seriously is a constant threat to essences, for in the economy of categories, whenever the value of accident changes, so, too, does the value of essence.[68]

In the context of art making, intentional, self-aware, creative apophenia can then be seen as a deliberate manipulation of ambiguity to explore the potential and the capacities of our brain and our senses. In the creation of art—as well as in the experiencing of art—the forging of interesting associations and connections between diverse, varied, and often unexpected subjects is enriching and fascinating. Once imbued with this spirit, we are constantly able to revise the world around us by discovering and revealing hitherto hidden patterns, which we can then join together aesthetically or conceptually.

At this point (and given the previous introduction of the idea of our internal reality as 'controlled hallucination'), it seems prudent to place a suitable amount of conceptual distance between the phenomena of apophenia and pareidolia with the more widely understood concept and experience of hallucination 'proper.' Clinical study on hallucination is not waning, by any means, abetted as it is by the increasingly dense pharmacopoeia inducing them, for example, substances like 5-Me0-DiPT that have gained their recreational popularity on the strength of "produc[ing] hallucinations associated with sounds more frequently than with vision."[69] However, the degree to which hallucinations match the experience of something like pareidolia seems low, especially in light of classic psycho-biographies like Paul Schreber's, in which he admits to auditory hallucinations being like a 'silent prayer' devoid of anything comparable to acoustic properties.[70] What can be said with more confidence is that hallucinations, whether we define them as 'just' self-referential over-interpretations of actual sensory perceptions or as actual pathological symptoms, seem intimately linked and, indeed, can be accounted for at a cognitive level for within the predictive brain framework, particularly as any hallucination involves unexpected changes in one's personal ability to "integrate incoming data with perceptual predictions."[71]

Now is perhaps an opportune moment to note Andy Clark's twist on the notion of perception as a 'controlled hallucination,' as it is useful for us to make such a distinction. Clark suggests that hallucination would be more appropriately described as a kind of 'uncontrolled perception.'[72] Typically, we think about experiences of hallucination as being something entirely different to normal perception, and we use the word to describe what is happening if we perceive something that isn't there. The point being made is that the same processes are shared between 'normal' perception and hallucination. Whether one is experiencing hallucinations, experiencing apophenia, or engaging in normal perception, the same processes are happening in the brain and only select aspects of the balance between these states has changed. This question of balance

---

[68] Hyde, *Trickster Makes This World*, 98.
[69] "DiPT the Auditory Hallucinogen. What Is It?" https://www.solutions-recovery.com/dipt-auditory-hallucinogen/. Accessed February 2, 2019.
[70] Schreber, *Memoirs of My Nervous Illness*.
[71] Griffin and Fletcher, "Predictive Processing, Source Monitoring, and Psychosis."
[72] Clark, *Surfing Uncertainty*; Clark, "Perception As Controlled Hallucination."

is key here. Extending Clark's suggestion that hallucination is a kind of uncontrolled perception, one could also more generally think of 'normal' perception as a kind of hallucination—just one in which perception is controlled, in the sense that predictions being made by the brain are kept in check by incoming sensory evidence from the external environment. As Seth has recently claimed, "we're all hallucinating all the time, including right now ... it's just that when we agree about our hallucinations, we call that 'reality.'"[73] Hallucinations and visual illusions, in accordance with this perspective, can be understood to be as real as anything else we might experience.

It has taken a good deal of time, since the formal delineation of hallucinations as a distinct category of perception, to arrive at this conclusion. Hallucinations only began to appear by name in pathological literature dating from 1830, though their auditory variety ('hearing voices' and related phenomena) is documented throughout all of recorded history, with the more historically resonating examples expertly relayed in Daniel Smith's historical survey *Muses, Madmen and Prophets*.[74] Interestingly, the definition of hallucination that Smith borrows from Esquirol seems, initially, to be a phenomenological synonym for pareidolia (to wit: "the inward conviction of a presently perceived sensation at a moment when no external object capable of arousing this sensation is within the field of [the] senses").[75] Fast forward to studies of phenomenology and affect taking place in the twenty-first century, and we have yet more definitions of hallucination which seem to refer to pareidolia: see Brian Massumi's characterization of hallucination as "the spontaneously creative addition of objects of perception that are not found preformed 'out there'" (of further interest here is Massumi's assertion that, in this process, our sensory apparatus "[gives] back more to reality than it is given").[76]

So, we have ample means with which to distinguish auditory pareidolia, and the sonic phantoms that they generate, from auditory hallucinations and their accompanying delusional behavior. Elsewhere, in his authoritative work on hallucinations, Jan Dirk Blom acknowledges auditory pareidolia as "music 'heard in' the background noise of engines, refrigerators, computer ventilators, and so on,"[77] while also providing further examples of auditory phenomena that do not count as hallucination (i.e., tinnitus and palinacousis,[78] though the former may in fact lead to bona fide hallucination should its sheer persistence begin to wreak havoc on the mental state of those suffering from it). It is also interesting to note that Blom includes in his dictionary of hallucination more paradoxical interpretations employing the term to describe 'normal' sense perception (as uncontrolled perception). He writes, "a consequent elaboration of this line of thought entails the view that all our percepts are hallucinations, and that sense

---

[73] Seth, "Anil Seth." See also Seth, "From Unconscious Inference to the Beholder's Share."
[74] Smith, *Muses, Madmen and Prophets*.
[75] Ibid.
[76] Massumi, *Parables for the Virtual*.
[77] Blom, *Hallucinations*.
[78] The latter is a rare form of auditory illusion reported in cases of temporal lobe epilepsy, defined as "sound persisting or repeating after the cessation of an auditory stimulation." See Fields et al., "Palinacousis, Palinacousis"; Bauman, *When Your Ears Ring!*

perceptions have the status of hallucinations that are merely modulated or restrained by the senses."[79] As noted by Blom, the French historian and critic Hippolyte Taine,[80] in his book *On Intelligence* written in 1872, concluded that external perception itself is an hallucination,

> We can now seize, with a general glance, on the process employed by Nature to create in us our first and principle sources of knowledge. In two words, she creates illusions and rectifications of illusion, hallucinations and repressions of hallucination.—On the one hand, with sensations and images combined in clusters according to laws we shall presently see, she constructs within us phantoms which we take for external objects, in most cases without being misled, since there are, in fact, external objects corresponding to them, sometimes, by mistake, since sometimes the corresponding external objects are not there; in this way she produces external perceptions which are true hallucinations, and hallucinations, strictly so called, which are false external perceptions.[81]

Hallucinations, as officially defined in medical dictionaries, are strong percepts of an object or event without any corresponding external stimulus being present. In other words, the brain perceives something happening that is not directly happening, and no one besides the recipient of these illusory percepts can detect them. Auditory hallucinations are actually surprisingly common, more so than hallucinations affecting any of the other sensory modalities. They can be broken down into two main subcategories; that is, hallucinations of the psychiatric kind and those of the non-psychiatric kind. Those of the psychiatric kind are characteristic clinical markers of many serious mental illnesses such as schizophrenia, bipolar disorder, post-traumatic stress disorder (PTSD), and so on, whilst those of the non-psychiatric kind are often experienced by those who have complaints of hearing loss, migraine, stress, insomnia, and so on. Such non-psychiatric auditory hallucinations have nothing to do with mental illness, and are rather symptomatic of defects and malfunctioning of the auditory circuits of the brain itself.[82] If phantom sounds are being heard, there are some clear indicators from a clinical perspective that signal as to whether they are psychiatric or non-psychiatric. Roughly speaking, people who experience psychiatric auditory hallucinations tend to hear meaningful *voices*—often clear and distinct— and people who experience non-psychiatric auditory hallucinations will more than often report hearing music or singing (that is, a repertoire of sonic phenomena more diversified than the sound of voices talking). Non-psychiatric auditory hallucinations are often reported to be more indistinct, like a vague radio broadcast or a TV program playing in another room, and mostly do not manifest as being 'meaningful' or personal (e.g., like the schizophrenic phenomenon of voices engaging directly with the hearer).

---

[79] Blom, *Hallucinations*, 442.
[80] Hippolyte Adolfe Taine (1828–1893) was a French critic and historian.
[81] Taine and Haye, *On Intelligence*, 248.
[82] Bauman, *Phantom Voices, Ethereal Music & Other Spooky Sounds*.

Such phantom perceptions can also be of singing or music so distinct that individual instruments and voices can be identified in great detail.

From the perspective of public perception, most auditory hallucinations are associated with mental illness, and so bring with them the fear or disdain attached to pathological conditions, whether justifiable or not. Auditory illusions experienced by the hard of hearing are widely reported, though many who suffer from them will not discuss their experiences due to such negative connotations. With common conditions outside the realm of psychopathology (hearing loss, for example), the brain no longer 'hears' what it used to hear and can make up for this loss by inventing all kinds of sounds and music that are not actually present. This can also be understood from the perspective of sensory deprivation. Psychedelic chemicals are well known as a cause of auditory hallucinations, but lesser known and more surprising are the countless numbers of prescriptive medications (although mostly unknown to the user) known to do the same.[83]

The focus of this volume is not on any kind of psychiatric auditory hallucination; instead our interests focus on the non-psychiatric, non-pathological, pareidolic, or apophenic type of creative and aesthetic auditory hallucinations or illusions that we are generally calling sonic phantoms. In contrast to those with a pathological or unintentional character, sonic phantoms, as understood here, are deliberate aural phenomena with creative and cultural aims, such as the generation of sonic-artistic and musical experiences. Despite the differences in their manifestations (by design, disease, or defect), however, there are many characteristics common to the experience of the non-psychiatric auditory hallucinations and sonic phantoms.

Constant background broadband noise (i.e., composed of many simultaneous frequencies) is a particularly common and interesting case, as it often takes on a musical quality for most listeners, irrespective of their mental or physiological condition. This is a form of auditory pareidolia, and one that enthusiasts of creative sound or music regularly experience, especially considering how a host of common domestic 'noise-generating' devices such as electric fans and air conditioners can provide the necessary conditions—broadband noise production, prolonged exposure—for such auditory pareidolia to manifest.

So, here is where we can reinforce the important distinction between auditory hallucinations and auditory pareidolia. True auditory hallucinations are those experienced with no apparent external stimulus. Auditory pareidolia—of which sonic phantoms would be an intentional sub-category—is the hearing of music or voices triggered by an unrelated background or collateral sound.

One key point to take away from all of this is that, in spite of their many non-overlapping features, auditory apophenia/pareidolia and clinically diagnosed auditory hallucinations are both manifestations of the same underlying neurological processes, which again reveal the constructive nature of perception: more accurately, they are

---

[83] Bauman, *Ototoxic Drugs Exposed*.

both processes that illustrate the degree to which our perceptual apparatus makes errors while *en route* to forming the most complete world picture it is capable of.

1.4 Creativity with Ambiguity

In discussing the evolution of animal and human communication, Gregory Bateson[84] once made a characteristically bold prediction: "[T]he logician's dream that men should communicate only by unambiguous digital signals has not come true, and is not likely to."[85] Even as our means of communicative technology continue to multiply, Bateson's decades-old statement has been borne out by the obstinate persistence of tools that operate on analog principles and whose operations are thus understood to proceed with a certain level of unpredictability or semi-randomness. The increasing inseparability of analog/carbon-based/offline reality from its digital/silicon-based/online derivative is proceeding quickly enough for a so-called 'ambient intelligence' (AmI) to emerge (read: an electronic environment whose 'electronic' character is invisible to the point of seeming fully integrated with its human constructors). Such an environment looks more and more to be a seamless hybrid of these supposed irreconcilable opposites. Even with completely solid-state sound recording at hand, many refuse this so-called 'sonic transparency' in favor of "musical pursuits in which distortion itself is part of the aesthetic."[86]

However, we do not even need to invoke technology to realize the truth of the situation: face-to-face, IRL ('in real life') communication largely unaided by technology also continues to defy the 'logician's dream,' as alluring communicative ambiguities proliferate in the form of new dialects and idiolects, themselves influencing fresh outbreaks of cultural novelty. One of the underlying central ideas underpinning the work with sonic phantoms, therefore, is that our surrounding aural 'normal' world is inherently ambiguous and uncertain, and requires interpretation. This ambiguity leads to specific kinds of experiences, which can be constructive and represent active engagement with such undefined sensory data, rather than a passive subjugation to it. Take auditory perception, as we normally understand it, through the act of listening: this seems like a passive process, but the truth is more complex and fascinating. By way of an analogy proposed by A. E. Denham, we might consider a 'wire-frame' illustration of a cube in which all of the outlines are composed of linked dots or points. When contemplating this image, Denham suggests

> It is almost impossible to see any individual point as just a point in space, independent of its role in forming a line; it is equally difficult to see each line so

---

[84] Gregory Bateson (1904–1980) was a multidisciplinary thinker most commonly associated with contributions to cybernetics and information theory, along with his collaborative efforts with his wife, the anthropologist Margaret Mead.
[85] Bateson, *Steps to an Ecology of Mind*.
[86] Poss, "Distortion Is Truth."

composed as just a line, independent of its role in forming a cube. Rather, one is compelled to perceive each 'under the aspect' of the whole—in this case under the aspect of a cube. Likewise, when a sound is heard as a tone, it is heard as a contribution to a larger pattern—a pattern within which its presence is heard as purposeful, as fulfilling a function.[87]

All of this serves as a potent reminder that, although our eardrums may be vibrating as a reflexive response to the mechanical pressure waves that eventually reach them, in order to eventually perceive something as a listening experience we must execute the neurological interpretational processes that give rise to perceptual patterns. The current literature suggests that predictive processing and coding models have of recent times become a dominant framework for studies in perception, including explorations of auditory perceptual phenomena. These models are also making use of a significant existing and detailed foundational research conducted over decades into auditory perception under the term 'auditory scene analysis' (hereinafter ASA), a term coined by cognitive scientist and psychologist Albert Bregman.[88]

ASA basically describes a fundamental skill of the auditory system enabling us to perceptually organize and identify sounds in our environment. Our auditory systems are often bombarded with a mixture of vibrating sound patterns originating from multiple different sources simultaneously and, when working well, our systems are able to disentangle sound mixtures. ASA is definable as our ability to segregate these sounds in our environment (phones ringing, people talking, cars driving, etc.) into separate perceptual units that Bregman terms 'auditory streams.'[89] By 'scene analysis' we mean how the auditory perceptual system analyzes and makes sense of a noisy and chaotic complex acoustic world, the auditory scene populated with sound sources that may 'overlap' in one or more qualitative features, by dividing it into individual coherent auditory components or meaningful events of interest. Relevant or meaningful auditory information can only be retrieved if the brain mechanisms succeed in decomposing this mixture into auditory streams, thus providing neural excitation patterns that maintain the integrity of the separate sound sources, allowing us to derive useful representations of reality from them. Bregman's concept of auditory streams plays the same role in auditory mental experience as the object does in visual experience; he reserves the word 'stream' for perceptual representation and the phrase 'acoustic event' or the word 'sound' for the physical cause.[90] It is useful in making the distinction between the acoustic information in the physical 'happening' and the perceptual representation (stream). The general scene analysis process therefore involves perceptually grouping, fusing, segregating the incoming streams of acoustic stimuli, in order to successfully detect meaning, structure, or pattern. This

---

[87] Denham, "The Moving Mirrors of Music."
[88] Denham and Winkler, "Predictive Coding in Auditory Perception."
[89] Auditory streams can be perceived as emanating from a single source and can be both real or fictional.
[90] Bregman, *Auditory Scene Analysis*, 10.

process of perceptually grouping incoming auditory events to form distinct auditory streams is referred to as 'auditory stream segregation.'[91]

The ASA problem was originally introduced by Colin Cherry in 1953 as 'the cocktail party problem'[92] before being later coined ASA by Bregman. This refers particularly to the capacity to follow one conversation when multiple competing conversations are taking place simultaneously in a noisy room. Cherry's work revealed that the ability to separate target sounds from background noise was dependent on many factors, including directionality of the sound, pitch, sex of the speaker and so on. Predictive processing comes into play here at this intersection between the incoming sensory information, and the top-down expectations or predictions (priors) about how the incoming sensory information is likely to be.[93] If we have very strong expectations about the incoming sensory input, then, given certain information, we can compensate for the missing information to the point of detecting phantom signals that are not actually present at all (we will be discussing this specific phenomenon in greater depth in chapter 4). If someone speaks your name quietly at a cocktail party, which is a noisy and ambiguous environment, the chances are that whilst you might have trouble hearing some parts of the conversation you will have no problem hearing your name clearly, and this again can be explained by your strong expectations about your name and what it sounds like. Anything that sounds remotely like your name will jump out of the composite noisy signal. Top-down expectation meets incoming sensory signals with a balance that is determined by your confidence level in either the incoming signal or the top-down predictions.

Bottom-up processing of the incoming signals, which has also been identified as 'attention-independent processing,'[94] given the lack of need for additional interpretation, is as much a part of our overall sonic experience. That experience is largely bound up with the extraction of zones of deliberate focus that we might call 'sonic objects' or 'auditory streams,' that is, acoustically similar sounds bound together by the hearer, from a sea of more attention-independent stimuli. Both the 'focused'/top-down and 'attention independent'/bottom-up models of processing, when acting in concert, are being carried out by our ASA mechanisms. When we succeed in the aforementioned act of identifying a single speaker in a crowded noisy room, we are successfully carrying out the task of auditory scene analysis. Casey O'Callaghan's amplification of this concept helps to achieve a better understanding of just what is involved in ASA:

> A set of grouping principles that involves assumptions about the objects of auditory perception enables us to associate correctly the *low pitch* with the *soft volume* and

---

[91] Bregman, *Auditory Scene Analysis*, 47.
[92] Cherry was a British cognitive scientist whose main contributions were in focused auditory attention, specifically the cocktail party problem. Cherry, Halle, and Jakobson, *Toward the Logical Description of Languages in Their Phonemic Aspect*; Cherry, *On Human Communication*.
[93] Predictive processing, the Bayesian brain, and hierarchical predictive coding, which for our purposes here are all more or less names for this same constructivist picture being described.
[94] Snyder et al., "Attention, Awareness, and the Perception of Auditory Scenes."

faraway location, and at the same time to group correctly the *high pitch* with the *loud volume* and *nearby location* without mixing things up into a garbled 'sound soup' of *high pitch, nearness, soft volume, low pitch, loud volume* and *distance* [all italics in the original].[95]

What O'Callaghan explains here seems so simple as to be taken for granted, yet there are moments when this sophisticated capability for partitioning sounds fails us. To be sure, adeptness with ASA requires more than just the inferential process of segregating sounds as they reach the ear: familiarity with certain environments plays a key role, and, as it is with interpreting visual messages within a sufficiently complex environment, there is always the possibility that receiving a coherent message means ignoring or overlooking some portion of the whole scene. We also cannot forget the helping hand that we receive from other sensory modes during this process, with visual information regularly contributing to how clearly we perceive acoustic qualities; and if this seems at all fantastic or improbable, we must remember we inhabit a world in which a pilot who anesthetized his rear end found that he could no longer orient himself via sight.[96] Also, though we may develop a talent over time for sonically navigating the 'cocktail party' with ease, things are less clear-cut within that category of sonic experience that untold millions would probably identify as their most memorable connection to the auditory world.

We are talking here, of course, about music (and it is not without reason that O'Callaghan illustrates the functioning of ASA with a 'day in the park' scenario rather than with a descriptive unfolding of a live music performance). The phenomenon of auditory streaming is fundamental to our understanding of melody and other kinds of polyphony. ASA can help us to make sense of complex acoustic mixtures such as that played by an orchestra, but in music scene analysis usually fails us in the sense that we cannot audibly distinguish every instrument in the orchestra. Meanwhile, auditory streams can point to both real and fictional sources. In natural environments, ASA helps us to build useful distinct perceptual representations of acoustic events such as a person talking, footsteps, a train in the distance, the wind, or rustling leaves. However, in music the aim is often to deliberately fool the auditory system into hearing fictional streams. In order for certain musical sounds to blend together, the listener must "defeat the scene-analysis processes that are trying to uncover the individual physical sources of sound."[97] Numerous compositional strategies can work to exploit this, such as the generation of pseudo-polyphony from a monophonic instrument. ASA and its universal principles of perception can be therefore be valuable compositional tools (and we will be examining many examples throughout this book relating precisely to this).

Of all the available illusions that we encounter when trying to parse audio streams, a disproportionate number seem to arise from encounters with composed music, and

---

[95] O'Callaghan, *Sounds*, 18.
[96] Pick and Walk, *Intersensory*.
[97] Bregman, *Auditory Scene Analysis*, 457.

manage to turn exercises in audition that seem simple 'on paper' into surprisingly complicated percepts. The reality of sonic illusion in music is especially true given our predilection for forming musical ensembles in which certain musical instruments or voices are represented multiple times to increase the richness of the overall sound. Determining the exact origin of each auditory component within, say, a church choir can prove immensely difficult without some sort of visible, gestural cues to guide us.

It is more challenging still to attempt this same task while listening to the frenetically swarming 'electric guitar orchestras' made famous by composers Glenn Branca[98] and Rhys Chatham.[99] Unlike the church choir, each voice in this type of choral arrangement is assumed to have an identical range of possible pitch values. This provides an example of sonic illusions that can arise from the Gestalt 'Law of Good Continuation,' which states that multiple sound components appearing close to each other in pitch, and, forming a regular 'pitch contour,' will be interpreted as originating from the same source and being part of the same event: a whole category of sonic illusions concerned with pitch differentials, including the scale illusion, the tritone paradox, and the octave illusion, bear out this thesis. The 'guitar orchestra' example is also a prime example of a central tenet of ASA—that there exists an inversely proportional, interdependent relationship between tempo and frequency among individual tones: that is to say, the quicker the tones follow one another, the smaller the amount of frequency separation at which they will segregate into individual streams. The hypnotic, blurred tremolo riffing of 'guitar orchestra' compositions such as Branca's numbered symphonies or Chatham's *An Angel Moves Too Fast to See* seem to bear this out (and the same could be said of the innumerable 'black metal' compositions that, despite being inspired by a significantly more over-the-top worldview than these 'serious' composers, nonetheless bear a technical kinship with them via the use of aggressively blurred 'tremolo riffing' and similar techniques).

Naturally, we should make sure not to ignore the penultimate 'fly in the ointment' of ASA that we could consider *musique concrète* to be: the compositional-aesthetic-philosophical realm created by Pierre Schaeffer took its name (along the way essentially inaugurating the musical concept of 'sampling') by inverting the classic music process and turning the abstract into the concrete, thus disrupting the process by which we typically make sense of the audible world and rendering it into musical compositions. By the time we arrive at something like *musique concrète*, in which the desired auditory state of 'schizophonia'[100] inspires a compositional method of *intentionally* frustrating our association of sounds with their exact sources (to say nothing of making these sounds themselves much more complex than what we might call 'pure tones'), then all bets are well and truly off. Far from being an accident, this was memorably confirmed

---

[98] Glenn Branca (1948–2018) was an American avant-garde composer and guitarist known for his use of volume, alternative guitar tunings, repetition, and droning.

[99] Rhys Chatham (1952–) is an American composer and multi-instrumentalist, best known for his 'guitar orchestra' compositions.

[100] Schizophonia is a term coined by Canadian composer/musicologist R. Murray Schafer to describe the way in which recording and playback technique results in the 'splitting' of a sound source from its natural or original environs. For a fuller discussion of the concept, see Schafer, *The Tuning of the World*.

as a deliberate strategy in many of Pierre Schaeffer's explicit pronouncements, such as the following:

> To *distinguish* an element (to hear it in itself, for the sake of its texture, its matter, its colour). To *repeat it:* repeat the same sonic fragment, there is not an event any more, there is music.[101]

So it should not go unmentioned that *musique concrète* and certain other musical forms, whether they call ASA by its name or not, aim precisely at confusing and upending this capacity for sonic discrimination, and use this as the jumping-off point for a paradigm in which sounds seem to have their own agency independent of us, and, furthermore, have a 'life' as objects of contemplation apart from their sources. The situation in which this parsing process fails, causing "sound components [to] fuse together, emerging perceptually as sound objects that are not actually present,"[102] has been embraced throughout the relatively short history of this form. In some cases, alternative forms of sound spatialization have been developed: timbral details within *musique concrète* and acousmatic music are often magnified to a degree uncommon in traditional music, and in the process create an illusion of 'different' spatial positions than what might normally be verifiable (this is, for example, the inspiration for the composition *Raumform* by Robert Platz,[103] as well as the psychoacoustic basis of the live immersive work by composer-performer Francisco López).[104] Moreover, well-trained acousmatic listeners and composers occasionally tend to complement the spatialization of sound by interpreting sounds as having dimensions normally associated with tactile physical objects (i.e., size, volume)—see especially composer Dennis Smalley's thoughts on the subject.[105]

Nevertheless, there is still an intriguing potential for perceptual illusions when we undergo the process of ASA; we do not need to adopt some other paradigm to defeat it before this can be achieved. As is the case with other sensory modes, ASA is susceptible to what can be termed 'bistability or multistability': typically used to describe a dynamic system in which there are two stable states of equilibrium, it can also apply to multiple points of mutually exclusive sensory information that can be alternated at will; for example, images in which multiple views can qualify as 'correct' interpretations. Similar in its perceptive effects to the 'Rabbit and Duck' image noted earlier, one of the most popularly recurring examples, which itself is a clear example of the so-called 'figure and ground' illusion, features the famous 'Rubin goblet' or 'vase' whose outlines are composed of two faces in profile facing towards one another: at any point, one of the two interpretations will dominate, with the 'faces' being seen as figures atop the ground/backdrop of the 'vase' or vice versa. Though such images strike us by their apparent novelty when being viewed for the first time, they are in

---

[101] Quoted in Palombini, "Machine Songs V."
[102] Scharine and Letowski, "Auditory Conflicts and Illusions."
[103] Platz and Wharton, "More Than Just Notes."
[104] López, "Against the Stage."
[105] Austin, "Sound Diffusion in Composition and Performance."

fact revealing fundamental facts of our perception that are also applied to our sense of hearing (and even to our olfactory sense, given the ability for different inputs to find their way to different nostrils). Understanding of bistability can contribute greatly to the understanding of those mechanisms contributing to perceptive stability as a whole: of equal importance, they reveal the way in which perception in general is an active, rather than passive, process.

What the above examples make abundantly clear is that the more ambiguous the perceptual 'scene' or sensory material, the more actively we need to work to make sense of it. However, while 'work' may imply serious physical or mental strain that leads to exhaustion rather than inspiration, that does not need to be the case for artistic practices, in which we are often rewarded aesthetically, intellectually, and emotionally after we have spent some time figuring out ambiguous scenes. In this context, the approach outlined here is an aesthetic one—in its widest sense—whereby we are perceptually open and willing to suspend certain critical faculties to achieve new creation by means of actively finding non-obvious patterns or meaning out of ambiguity.

In terms of working creatively with sound, this active effort results in a heightening of the listener's auditory experience as it, making good on the aforementioned theory of the 'beholder's share,' actively participates in the perceptual process. When working with ambiguous sonic textures, for example, an aesthetic mode of listening (again, in its

Figure 1.7 'White's illusion.' The two vertical columns with alternating grey and white rectangles appear to be different but are in fact exactly the same tone. This illustrates the fact that target luminance can elicit different perceptions of brightness in different contexts. Image (public domain) by Lockal—Wikipedia Commons (https://commons.wikimedia.org/wiki/File:White_illusion.svg).[106]

---

[106] White, "The Effect of the Nature of the Surround on the Perceived Lightness of Grey Bars Within Square-Wave Test Gratings."

broadest sense) greatly influences which patterns and details emerge into awareness, and to what extent they manage to do so. That the musical experience is dramatically impacted by the degree to which listeners' interaction is 'active' could, of course, be a general statement applicable to most music pieces, but, in the case of deliberate and meaningful ambiguity, such a degree of involvement takes center stage as a necessary prerequisite for the essence of this particular type of musical and sonic experience to manifest.

I find this development of intentional multiple listening perspectives particularly appealing as a compositional aesthetic. As I believe will become evident from the different music pieces that will be discussed and examined over the course of this volume, this specific type of ambiguity pervades not only all of my work, but a good deal of music and sonic art that has since come to be recognized as groundbreaking or generative of new cultural shifts. Some sonic phantom patterns arise as a consequence of the deployment of multiple and simultaneous potential auditory 'images' or 'scenes,' which will either reveal themselves or perceptually dissipate depending on the specific listening circumstances. In my compositions I have intentionally worked with our propensity to experience auditory pareidolia when faced with ambiguity. I play with sonic materials that are either naturally ambiguous or have been 'composed' to attain this quality, in order to exploit the potential for auditory pareidolia to manifest, bringing with it the phantasmatic presence (this has resulted in works where, for example, within formless, abstract, or noisy textures we begin to hear voices or imagine voice-like forms or entities). I have developed and adopted many diverse, and somewhat unorthodox, approaches and techniques to achieve this aim, which I will elaborate on below with a number of compositional projects.

1.5 Ambiguity as Intentional Practice

> *… many effects depend on the readiness of the listener to perceive them, but these lead us on into the grey area of faith and hope where creative illusion and the deception of our senses through natural phenomena lie closely inter-twined. This is the very frontier at which music is born.*
>
> —Robert Platz[107]

The practice of artful deceit is one that has, perhaps, been taken for granted in an era typified by phenomena such as false online identities, and the recent emergent trend in artificial intelligence and the human image synthesis known as the 'deep fake.' In such a deceit-saturated infosphere, it seems almost quaint to recall a time in which, say, the marshaling of Surrealist painters to design military camouflage was a strikingly anomalous example of placing the archaic knowledge of illusions in the service of techno-scientific endeavors. However, in spite of the current omnipresence

---

[107] Platz and Wharton, "More Than Just Notes."

of deceit and subterfuge, and the negative consequences this can have upon all forms of interpersonal communication, we should not lose sight of the ways in which the archetypal 'trickster' figure has been a positive force for the evolution of creativity (and not least the evolution of sonic creativity). With terms such as 'confidence trickster' not losing any currency, and painting an altogether different picture of the trickster as a petty sociopath, it is useful to delineate exactly how that is the case. Naturally, it is also essential to illuminate tricksters in their creative aspect, since the enlightening ambiguity that they embody is part and parcel of our shared sonic reality.

One of the best tutelary figures to demonstrate a more creative or constructive type of trickster-ism is the character of the shaman, with shamanism itself being described broadly by Jean-Jacques Nattiez as "the entire set of practices and rituals done by all members of an animistic society."[108] Many such societies have existed: though the name 'shaman' can be traced most directly to the Tungus language of Siberia, it has become a global shorthand for an archaic fusion of healer, mystic, and priest roles, which in combination provide one of the most enduring representations of the trickster strain of creativity. In the myriad ancient societies that had shamanism at the center of their spiritual life, the shaman came to be synonymous with this archetype, to the point where that role, and the more characteristically benign attributes of the shaman as a folk healer and 'exerter of influence' upon other biological organisms, may seem to conflict with the more devious trickster designation. Yet, as anthropologist Piers Vitebsky suggests, these functions are all part of the same toolkit; the trickster-ism endemic to this culture is not used (solely) for self aggrandizing purposes, but as a means of placing the shaman on an equal footing with deities whose intentions towards their earthly counterparts are not entirely benevolent:

> Being a trickster is an essential strand in the make-up of a shaman, who must change form to fight and outwit obstructive spirits. Primeval shamans used trickery to capture the sun so as to give people daylight, or stole the secrets of fire, hunting or agriculture from jealous spirits.[109]

In this context, I frequently use the phrase 'in-between moments' to describe the shape-shifting world that I gravitate towards. I am drawn to the creative potency of such liminal spaces (from *limen*, Latin for threshold), which are inherently ambiguous, lying *betwixt and between* one thing and another. Deeply connected with this concept of liminality, as elaborated first by ethnographer Arnold van Gennep and later by anthropologist Victor Turner,[110] and latent throughout my artistic practice, is the primordial figure of the shape-shifting, confusion-sowing, gleefully equivocal trickster. By many accounts, the trickster, aptly described as the 'God of the in-between,' prefers to hang out in doorways, in the in-between spaces, then in the rooms themselves.[111]

---

[108] Nattiez, "Inuit Throat-Games and Siberian Throat Singing." See also Cardeña and Beard, "Truthful Trickery."
[109] Vitebsky, *The Shaman*.
[110] Turner, *The Ritual Process*.
[111] Hyde, *Trickster Makes This World*.

Mircea Eliade's own landmark studies of Siberian and Central Asian shamanism also confirm this notion in many salient respects, given that numerous traditions envision the character of the shaman as a psychopomp having the ability to walk between worlds, or an individual who acts as a conciliatory force between the life-worlds of animals and humans (indeed, the Buryat 'origin story' of the first shaman claims that such a figure arose from the coupling of a male eagle and a human female). Such concepts linked with the trickster figure, as well as the broader psycho-spiritual condition of boundary crossing, boundary loss, and blurring of distinctions, provide a framework with which I can relate the ambiguous shifting of 'figure' and 'ground' (to use the classic visual analogy) in my own work. The conceptualization of the trickster as a revealer of new worlds also has plenty of implications for the continued development of music and sonic art.

It is unsurprising, then, that this approach should become a regular feature of various strains of performance culture in which performance itself is not the final product, but a process of mutual, or reciprocal, self-expression shared between performer and audience. When the Northern Irish action artist André Stitt at once refers to himself as an avatar of the trickster figure and claims that his 'core ethic' is "wanting to strategically re-structure the nature of reality,"[112] he hints at a strong modern tradition of artists self-identifying with the trickster archetype; particularly artists who are inhabiting the interstitial spaces between officially or institutionally recognized creative media or those whose practice involves testing the boundaries of 'acceptable' discourse and conduct (and, in keeping with the blurring of distinctions already in play, these two categories of activity are not mutually exclusive either). Though an iconoclastic or antinomian stance is not a prerequisite for understanding the nature of sonic illusions in particular, it could be argued that understanding reality as being in constant flux can only aid in extracting greater meaning from perceptual ambiguity.

I feel that the most substantial creative discoveries of my working process—as it probably is the case for many other artists—often occur in-between the intentional and the accidental. Most of my working process is initially exploratory in nature and driven to a large degree by being attentive to happenstance and accidental discovery. In his book *Trickster Makes This World*, Hyde uses the elegant metaphorical phrase "a net to catch contingency" to refer to the process in which, through happenstance, the happy accident of creation is born to those with the insight to recognize it.[113] Leonardo da Vinci famously wrote a passage in his *Treatise about Painting* on such accidental discovery as a way to "awaken the genius," which is worth quoting in full:

> I will not forget to insert into these rules, a new theoretical invention for knowledge's sake, which, although it seems of little import and good for a laugh, is nonetheless, of great utility in bringing out the creativity in some of these inventions. This is the case if you cast your glance on any walls dirty with such stains or walls made up of rock formations of different types. If you have to invent some scenes, you will be able to discover them there in diverse forms, in diverse landscapes, adorned with

---

[112] Stitt et al., *Small Time Life*.
[113] Hyde, *Trickster Makes This World*.

mountains, rivers, rocks, trees, extensive plains, valleys, and hills. You can even see different battle scenes and movements made up of unusual figures, faces with strange expressions, and myriad things which you can transform into a complete and proper form constituting part of similar walls and rocks. These are like the sound of bells, in whose tolling, you hear names and words that your imagination conjures up.

Don't underestimate this idea of mine, which calls to mind that it would not be too much of an effort to pause sometimes to look into these stains on walls, the ashes from the fire, the clouds, the mud, or other similar places. If these are well contemplated, you will find fantastic inventions that awaken the genius of the painter to new inventions, such as compositions of battles, animals, and men, as well as diverse composition of landscapes, and monstrous things, as devils and the like. These will do you well because they will awaken genius with this jumble of things. But, first you must know the components of all those groups of things you wish to represent, such as the members of the animal kingdom, as well as the components of the countryside, such as rocks, plants and similar things...[114]

As part of a long lineage that would include Protogenes's and Leonardo da Vinci's attention to accidental stains for inspiration in painting,[115] as well as the old German children's game 'Blotto' (klecksographie)[116] and Rorschach's inkblot psychological test,[117] 'Blotscape' is a term created by Janson to describe the visual inkblot method of eighteenth-century painter and 'blotmaster' Alexander Cozens:

[T]he British landscape painter and drawing teacher Alexander Cozens... published an illustrated treatise entitled A New Method of Assisting the Invention in Drawing Original Compositions of Landscape. It describes "a mechanical method... to draw forth the ideas" of artists, which consists of making casual and largely accidental inkblots on paper with a brush, to serve as a store of compositional suggestions.[118]

Cozens's method, unorthodox for the time, was an innovation in the realm of landscape painting rather than the much more abstract forms of the early twentieth-century's avant-garde, which have more in common visually with Hermann Rorschach's well-known tool for diagnosis. These, too, bear mentioning: a series such as Hans Arp's illustrations for his Dada comrade Tristan Tzara's *Vingt-cinq poèmes* (which actually pre-date, by a few years, publication of the hand-drawn 'blots' which would earn Rorschach his lasting fame) displays remarkably well how apophenia can take hold in viewers, and how the determination to subjectivize an 'abstract' image can instill a

---

[114] Da Vinci, *"Trattato Della Pittura" Di Leonardo Da Vinci*.
[115] See Gamboni, "Stumbling Over and Upon Art"; Gamboni, *Potential Images*.
[116] 'Klecksography' was the art of making images from inkblots. The work was pioneered by Justinus Kerner.
[117] The Rorschach inkblot test is a projective psychological test using ten inkblots printed on cards created in 1921 with the publication of *Psychodiagnostik* by Hermann Rorschach. Rorschach, *Psychodiagnostik*.
[118] See H. W. Janson, "Chance Images," in Wiener, *Dictionary of the History of Ideas*.

trance-like focus in the most committed of those viewers. Arp's thickly inked figures, as incoherent as they may initially seem to be, have an elasticity and symmetry of form that makes it difficult to ascribe anything other than 'human' characteristics to them, and which thus work to draw out highly personalized narratives from the viewer in the same way as 'Rorschach blots' later would.[119]

Following from the principles of active viewership that animate such phenomena, by analogy I use the term 'sonic blotscape' to refer to a hands-on, practical experimentation with sound, with intentional ambiguity at its core, which typically leads to accidental yet fruitful sonic discoveries. A sonic blotscape would accordingly be an equivalent territory of apparently formless, or seemingly unrelated, sonic geography, which would however be full of potentially countless auditory versions of the 'chance image' with an ever-changing multistable quality. The sonic blotscape functions for me in a similar sense to Hyde's 'net to catch contingency'. It activates sonic phantoms and induces them to manifest in the listening imagination. It functions to inspire, project, and suggest 'meaningful' forms as well as encouraging an intriguing and heightened phantasmatic listening experience. For the apophenic, pattern-forming, open mind, this is fertile terrain for the induction of sonic phantoms.

As we can 'compose' visual landscapes out of wall stains or inkblots, so we can discover a plethora of meaningful patterns or voices from formless ambiguous noise. As a form of de facto definition, I will describe in the following chapters the specific and practical ways in which I have attempted to create such a perceptual generative territory with different sound materials.

Figure 1.8 Examples of some of my own digitally generated 'blotscapes'. Image by Barbara Ellison, 2019.

[119] The novelist Victor Hugo (1802–1885) was an avid blotter, as can be seen in his very sophisticated and elaborated blot-inspired pen-and-ink drawings.

Figure 1.9 *Something Bumped against a Wall at Work and Made a Painting of a Snowy Town.* Accidentally produced pattern of wall paint photographed by Reddit user BonerificNoodles (with kind permission; first posted on Reddit in 2017).

Figure 1.10 *Snowy Town.* Previous figure framed and retitled. "Unknown artist," paintscrape and steel, 2017. Posted by Reddit user 'BonerificNoodles' (with kind permission; first posted on Reddit in 2017).

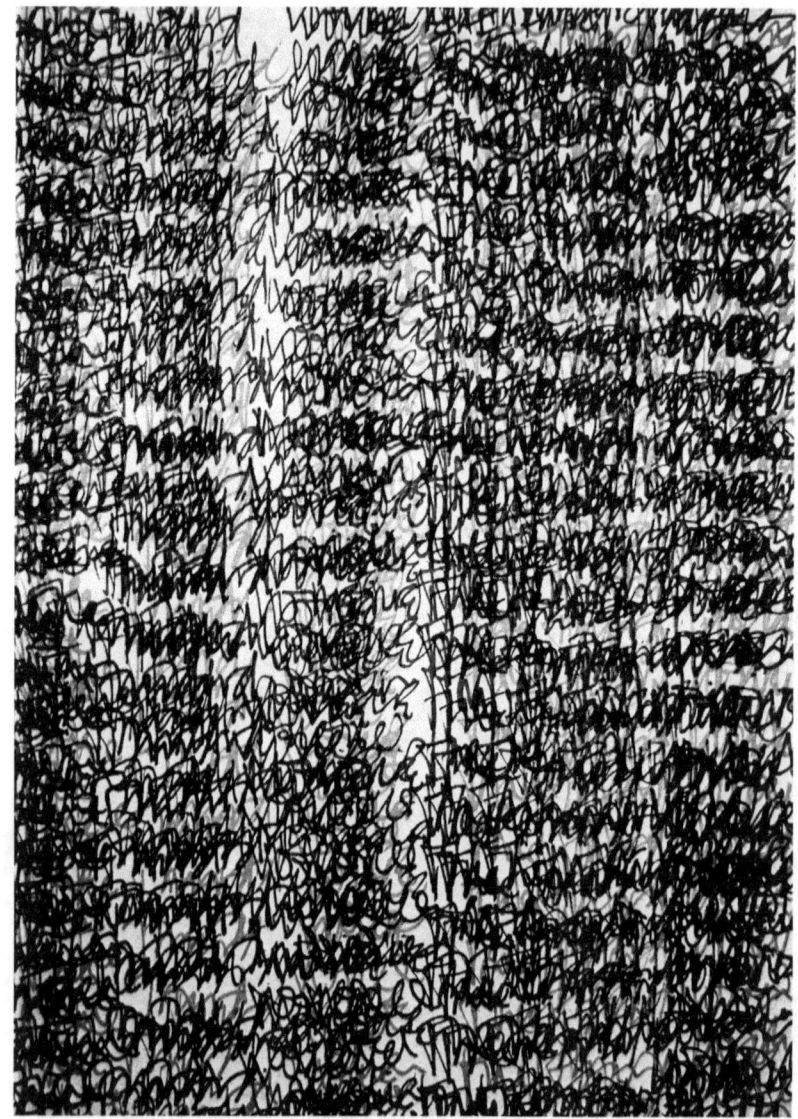

Figure 2.1 *The Third Voice*. Original writing-drawing artwork created for the poster of the performance of *Vocal Phantoms* at Störung #9 Festival (Barcelona, Spain, 2014). Image by Barbara Ellison, 2013.

2

Inducing the Phantasmatic

2.1 Compositional Phantasmatic Strategies and Techniques

In my personal compositional practice, I have used a variety of strategies and techniques to give rise to sonic phantoms and make them perceivable as an integral part of the compositions. I have developed these strategies and techniques as compositional tools, in very different instrumental or material realms and scenarios, while researching and experimenting with dramatically diverse sonic materials, as well as during the compositional work itself.

With the aim of providing a generally applicable overview, I have tried to identify what I consider to be the truly substantial, if not fundamental, processes that inform those strategies and techniques. When analyzed this way, one eventually realizes that the essence of such particular creative tools is not unique to a narrow band of experimental music practices: it is naturally widespread and it also extends beyond the domain of music, into other artistic endeavors and into other realms of self-inquiry, such as the ritual/spiritual. These essential processes are outlined, and put into perspective within a larger sonic creative context, in the following sections. The details about the implementation of particular strategies or techniques through specific instruments, devices, and ideas are described in each of the corresponding chapters covering the different compositional series.

2.1.1 Repetition

Repetition changes nothing in the object repeated, but does change something in the mind that contemplates it.[1] Throughout recorded history and across cultures, repetition is widely found as one of the essential defining features of music and other creative sonic practices. Repetition as a general strategy can be used to give rise to a multiplicity of remarkable perceptual, emotional, aesthetic, or semantic effects. It is a traditional mechanism of ritualization and trance induction, and to some extent it appears as such in my work. For example, in both my *Harp Phantoms* and *Drawing Phantoms* compositional series (described in detail in chapters 3 and 4) there are explicitly ritualized actions in which the performer, under the right conditions,

---

[1] Restating Hume's famous thesis, Deleuze, *Difference and Repetition*, 90.

might reach a trance-like state by virtue of repetitive manual actions on the musical instrument or by systematic repetitive circular motion on an amplified drawing surface.

One of the consequences of this type of repetition is the 'semantization' of sonic elements that were initially meaningless and would remain so in the absence of repetition—a clear example of apophenia. This semantization commonly produces 'speaking' sonic phantoms, apparent voices with typically brief meaningful 'messages' (repetition of short fragments of recordings is one of the techniques of EVP, or the presumed recording of the voices of the dead: see chapter 4).

Interestingly, repetition can also produce precisely the opposite effect, that is, the dissipation of meaning by 'semantic satiation,' the common phenomenon we experience if we insistently repeat a familiar word until we realize it has become semantically meaningless. This is the case in many aspects of my compositional series *Vocal Phantoms*, where extreme repetition of real and synthesized speech first dissipates the original sensation and then ends up transforming into an abundance of remarkably specific sonic illusions that are completely different for different listeners (referring back to apophenia, these are typically non-semantic but refer to imaginary objects).

These types of sonic phantoms produced by repetition are characteristically and naturally 'musical' (in a traditional sense) for most listeners, because of the fundamental prominence of rhythm and repeated structures in most music. The generation of 'musicality' is virtually an instantaneous phenomenon in all forms of 'looped music' based on non-musical sounds (e.g., audio artists such as Croiners,[2] Manon Anne Gillis,[3] [The User],[4] different strands of electronica; 'locked-grooves' records). It is equally the case even for standard film materials with diegetic sound when subjected to extreme repetition, as in the work of visual artist Martin Arnold.[5]

Repetition as a mechanism also reveals additional, inherent layers of alien or occluded musicality within music or musical sound, by perceptually deconstructing the original sounds into separate polyphonic, rhythmic, or harmonic layers that become audible only by means of repetition (e.g., as a prototypical feature of Western

---

[2] Croiners is the recording alias of James Levine, whose sound work is regularly built from looped phrases combining synthetic and natural sources.
[3] Manon Anne Gillis (also referred to simply as Anne Gillis) is a French performer, installation artist, and sound artist active since the early 1980s, who occasionally incorporates the materials of the recording process (magnetic tape, etc.) into sculptural/art objects.
[4] [The User] is a duo of sound artists Thomas McIntosh and Emmanuel Madan; their creative focus in the past has been upon recently obsolete technologies (see, e.g., their *Symphony for Dot Matrix Printers* [Staalplaat, 2002], which uses the titular device as the only sound source).
[5] Martin Arnold is an experimental filmmaker known for his obsessive deconstruction of found footage.

contemporary musical 'minimalism'; or, even more specifically, in the work of artists as diverse as Maryanne Amacher,[6] John Oswald,[7] or Oval.[8]

Such was the case as well, if we refer to the Western musical realm, of Satie's[9] *Vexations*, whose single-sheet, thirteen-bar piano score calls for it to be repeated 840 times.[10] Undoubtedly, where Western music is concerned, this type of intense repetition has been enabled by technological development. Maryanne Amacher, in detailing her personal experiences with electronic music, posited that electronic instruments were valuable not necessarily because of the novel sonic qualities that these instruments promised, but because they allowed her to confront new psychoacoustic effects before initiating an audience into the same:

> It was possible for me to make these discoveries *experientially* before considering how I would develop them musically...I had unrestricted time to observe what I was experiencing which would be severely limited if working with instrumentalists.[11]

This is just one example among many that shows how advances in music recording and playback technology have undoubtedly made it easier, and more attractive, for music of an intensely repetitive nature to be performed (even if this is a double-edged sword that regularly invites comments about the supposedly perfunctory, 'just press play and stand back' nature of some of this music).

One of the simplest means of conjuring repetition via technology, as is known to anyone with consumer-grade audio multi-tracking software, is the creation of seamless looped phrases. The use of the term 'looping' as an audio technique may seem crude and limited to musicians and sonic artists striving to make works characterized by continual differentiation, though the fascination with this technique may stem from

---

[6] Maryanne Amacher (1938–2009) was an installation artist and composer whose exploration of distortion product acoustic emissions (to be discussed later in this volume) remains fairly unique among sonic researchers.

[7] John Oswald (1953–) is a sound artist and originator of the term 'plunderphonics,' which refers to the liberal use of pre-recorded material (recognizable or no) in order to create new compositions. Oswald's work in this area has also been influential upon discussions about the legality of sampling-based composition.

[8] Oval is the production alias of German sound artist Markus Popp (formerly with Frank Metzger and Sebastian Oschatz). Their collective efforts are one example of so-called 'glitch' music, i.e., the introduction of chance elements into digital music by means of physical alterations (i.e., damaged compact discs) or semi-randomized computer processes.

[9] Erik Satie (1866–1925), French composer of works such as *Trois Gymnopédies* for piano and the ballet *Parade*, pioneered a unique minimalist form that contributed significantly to later varieties of 'ambient' music.

[10] Mary Davis explains the significance of the title by claiming the "vexation" arose not only from its "repetitious nature" but also "its persistent enharmonic spellings, atonal harmonies and asymmetrical phrase structures, all of which undercut efforts to retain the music in memory." Davis, *Erik Satie*.

[11] Amacher, "Psychoacoustic Phenomena in Musical Composition."

more sublime reasoning than looping's 'ease of use' alone. Michael Gendreau's[12] introductory explanation in his book *Parataxes*, for one, suggests deeper motivations for the seductiveness of the looped phrase:

> So many different elements of physical reality, and our consciousness of it, play into the perception of loops and their effects. They involve human memory, sometimes they are perceived simply as rhythmic motives and involve time in this respect, [at other times] they involve the perception of time in that it is easy to create time flow illusions with a loop.[13]

The mention of consciousness here brings us to another more deeply embedded type of 'loop,' the neural loops or feedback paths in the brain as proposed originally by Rafael Lorente de Nó. His discovery of this neural mechanism, later developed further by Donald Hebb, went some way towards explaining why looped sonic material can maintain the hold on us that it does, that is, 'synaptic strength' is increased with repeated activation of the loop. Explained in more detail, "if the axon of an 'input' neuron is near enough to excite a target neuron, and if it persistently takes part in firing the target neuron, some growth processes take place in one or both cells to increase the efficiency of the input neuron's stimulation."[14] Seen in this light, loops of audio material can be understood as manifestations of processes intrinsic to our entire psychological development.

One of those processes goes by the name of 'auditory driving,' which involves our brainwave rhythms and bodily functions (such as breathing or cardiac rhythms) entraining to external rhythmic stimuli; that is to say, they become synchronized to the external auditory rhythm. It is for this reason that looping and repetition are such powerful trance-inducing mechanisms. The repetition of mantras, the spinning of a dervish, the chanting and drumming of shamans, all induce different affective varieties of the trance phenomenon by limiting our attention and overloading our mind with repeated thoughts. The purposes behind each experience may be different, and the results will be certainly different at a subjective level, but they could all be considered trances. The looping of awareness is fundamental, and from practical experimentation it appears that the looping/repetition strategy alone is a vital prerequisite to induce a dissociated state. When cognitive loops are intensely repeated, eventually a type of mental dissociation inevitably takes place.

Returning briefly to Gendreau's statement above: one of the more interesting innovations allowing for the type of multivalent looping he mentions does not require modification to an existing musical instrument, but rather the elevation of audio storage media to an 'instrumental' status. The 'circularity' of the loop is very easy to

---

[12] Michael Gendreau is a Californian sound artist, formerly of the group Crawling with Tarts and an engineer in acoustical physics.
[13] Gendreau, *Parataxes*.
[14] Milner, "The Mind and Donald O. Hebb."

associate with sonic storage media whose physical design implies as much, and so, unsurprisingly, vinyl records have long been at the center of this musico-historical trend. Pierre Schaeffer's *courante*—a movement within a larger suite of concrete music—employed the "looping-groove technique...an effect similar to that of a scratched record,"[15] in such a way that it would liberate individual sounds from an enclosing context and make them comparable to words "in the state of liberty they enjoy in a dictionary...isolated not only from dramatic or anecdotal context, but also from musical context."[16]

This brings us to one of many intriguing ironies of this sonic adventure, in this case the use of a relatively minuscule unit of sound data, the 'loop,' to create immersive environments with a large time-scale. Unintentionally following on from Schaeffer's aforementioned explorations (as popular electronic dance music forms have time and again), exploitation of the 'locked groove' feature on vinyl records has been one means of creating such environments. Initially designed for the purpose of keeping the record player's tone arm 'locked' into the disc, and thus preventing it from causing audible annoyances and potential damage by straying into the run-off area, this quirk of record-cutting technology was eventually adapted so that locked grooves could occur *anywhere* on the surface of a record. This has led to numerous sound pieces of short duration which could be looped as many times as external circumstances would allow, an innovation that has occasionally been utilized within the culture of techno DJ-ing. In sub-genres such as so-called 'minimal techno,' a decided emphasis exists on enforced repetition as a means towards advanced cognitive and emotional awareness, perhaps brought on by the quasi-meditative, paradoxical practice of dancing more and more energetically merely to stay in place (and there are certainly practitioners of this music who see themselves, in the face of strongly opposing viewpoints, as carrying on a shamanic tradition by engaging in such).[17] In order to help maintain the inhuman precision needed by the DJ to make smooth segues between individual songs, locked grooves have become a sort of close relative of the 'tool' track, this being a 12-inch record side or digital audio file that simply reprises a particular looped motif of its 'parent' track for the duration of that track.

Among those artists who have been more explicitly associated with experimentalism in the public imagination, there have been a handful of albums whose content either heavily featured or was completely comprised of repetitive 'locked groove' pieces: Lee Ranaldo's 1987 record *From Here To Infinity*, for example, features each of the

---

[15] Palombini, "Pierre Schaeffer, 1953."

[16] Ibid.

[17] One such skeptical view of techno music's 'shamanic' ability comes from a distrust of its ability to truly interact with an environment as more archaic forms of shamanic music might. This is exemplified by the protest of the late percussionist/sound artist Z'ev, who states "You can't do the dance with electric music...music was/is used in pre-technical cultures to effect changes in the environment; electric music can only invade its immediate environment." Quoted in liner notes from Z'ev, "1968–1990: One Foot In The Grave."

otherwise brief compositions ending in a locked groove (interestingly, Ranaldo's flagship band Sonic Youth[18] had previously featured a 'lock groove' on their *EVOL* album, whose duration was listed as ∞ on the album liner notes). Elsewhere, the *Pagan Muzak* 7-inch release from Boyd Rice's[19] NON, while ostensibly a vinyl 'single,' utilized a similar technique to expand small fragments of jarring concrete sound into more monolithic structures (as an added bonus, the record's original pressing featured multiple center holes drilled into it and was recommended to be 'played at any speed,' implicitly challenging the rest of the then-novel Industrial Music subculture to tarry in the realm of conceptual audio releases).[20]

Perhaps most ambitious of all was RRRecords' series of compilation records which featured 100, 500, and 1,000 locked grooves respectively, a feat that caused plenty of exasperation among listeners who (though presumably knowing what they were getting into) nonetheless found it difficult to differentiate the individual tracks on these records: with hundreds of distinct tracks crammed onto one side of an LP, and the requirement for the needle to be moved by hand each time listeners wanted to advance in the program, the slightest adjustments to the tone arm position could result in listeners inadvertently skipping dozens of tracks and, consequently, having no idea whether or not they had fully absorbed the contents of the album. Noise artist Ron Lessard, the label owner responsible for curating and releasing the series, let slip some of the perverse humor at the heart of the project when apologizing for possible mastering issues that were causing the needle to jump from one groove to the next: "if this copy should track only 990 lock-grooves, then please accept my apologies for the inconvenience."[21]

Such devices, as well as mechanical repetition in general, have also added an extra dimension to the works of poets working in the 'text-sound' or sound poetry field, a creative realm that may be best defined in practitioner Sten Hanson's words when he describes the form as "the combination of the exactness of literature and the time manipulation of music."[22] Dean Suzuki suggests much the same, with greater humility, when claiming "sound poetry has always been a fringe art and at the periphery of the world of poetry."[23] Whatever one's assessment of its ultimate relevance and art-historical status, even a cursory overview of the sound poetry canon reveals it to be highly demonstrative of phantom sonic effects. Its ongoing relationship with repetition is just one of the more notable.

---

[18] Sonic Youth was a New York-based independent rock group, active from the early 1980s until the mid-2010s, widely recognized for incorporation of dissonance and alternate guitar tunings into compositions with more traditional 'rock' structures.

[19] Boyd Rice is an American conceptual artist and musician whose valuation of the prank as a viable art form, coupled with his experiments with pure noise, played a starring role in shaping the 'industrial culture' aesthetic of the 1970s–1980s and beyond.

[20] For more detailed info on such topics, see Bailey, *Unofficial Release*.

[21] Various, RRR-1000 Lock Grooves, https://www.discogs.com/Various-RRR-1000-Lock-Grooves/release/1973542. Accessed November 21, 2018.

[22] Quoted in Wendt, "Vocal Neighborhoods."

[23] Suzuki, "A Polypoetical Collision."

Aram Saroyan, the minimalist poet infamous for controversy-courting, monolexical pieces such as *lighght*, composed a similarly atomized piece of poetry, *Crickets*, which was featured in the run-off grooves of *10 + 2 = 12 (American Sound Pieces)*, the first significant compilation album of American sound poetry. As might be expected, the piece consisted solely of its title being repeated ad infinitum, causing a sort of dissociative process to occur wherein the spoken word 'crickets' became separate from that which the phonemes represented, or the sonic qualities of the piece became more an object of contemplation than the titular insect. Interestingly, the opposite side of the same album featured another piece, Anthony Gnazzo's *Population Explosion*, which took a virtually identical sonic approach to Saroyan's piece, this time using numbing repetitions of the word 'bang' (this being the onomatopoetic sound associated with an explosion) to seemingly implicate human reproduction itself as an action fraught with danger.

Upon closer examination, many of the fundamental figures in text-sound composition have experimented with techniques of repetition of single words or brief phases which, if not already altered by techniques such as progressively dense multi-track layering (see, e.g., Charles Amirkhanian's[24] *Mushrooms (for John Cage)* on the Giorno Poetry Systems LP *Totally Corrupt*), call upon the mind's natural inclination towards 'order-seeking' in order to produce similarly disorienting effects. In Amirkhanian's own reckoning, this process is strikingly similar to the already explored uses of apophenia in sonic art: his "abstract objectification of words into sound" is described as follows: "repeated figures... sustained for short periods of time before giving way to other variations on similar themes. This is pursued in several voices polyphonically, creating kaleidoscopic effects within his typically Webernesque time frames."[25]

This is a technique that has fit remarkably well into the repertoire of other artists who have found sound pieces to be an ideal or necessary supplement to their work in other media. One such case is Bruce Nauman:[26] perhaps more widely known for his groundbreaking efforts in video art, the earliest of which involved the artist carrying out apparently meaningless actions for excruciatingly long periods of time, Nauman also found creative uses for intense phonic repetition. A 2005 exhibition of sonic extracts from his pieces in other media, *Raw Materials* at the Tate Modern gallery, was the fruit of this. Individual pieces within the installation such as *Think Think Think*, *OK OK OK*, *Work Work*, and *No No No No* (the precise contents of which can all be easily intuited once these titles are known) all underscored Nauman's central concerns of "[exploring] the poetics of confusion, anxiety, boredom, entrapment and failure."[27] Meanwhile, the

---

[24] Charles Amirkhanian (1945–) is an American 'text-sound' artist, whose efforts in that area include composition of his own work as well as significant collaborative/educational efforts to make listeners aware of the internationally distributed nature of this form.

[25] Ubuweb: Sound, Charles Amirkhanian, http://www.ubu.com/sound/amir.html. Accessed November 21, 2018.

[26] Bruce Nauman (1941–) is an American multimedia artist particularly known for his contributions to the establishment of the video art genre in the 1960s.

[27] Rondeau, "Clown Torture, 1987 by Bruce Nauman."

artist's self-limitation to exclusively audio content had a more specific aim of "allowing almost random associations to appear,"[28] and simultaneously of "reinforc[ing] the musicality and the emotional content, rather than the intellectual content."[29]

While a wealth of material has already been suggested here for further examination, it is the speech-based compositions (alternately 'speech-based exercises,' for those who bristle at 'compositional' qualities being ascribed to serial utterances of a single word) that might be most intriguing in terms of their phantasmatic quality. This is partially due to the heightened emotional resonance associated with the 'humanity' of discernibly vocal material; something seized upon by composer Jacob Ter Veldhuis for its "wider *tessitura* or range, more varied topography, and melodic potential." With this last point in mind, the New York-based composer Scott Johnson also suggests a psychoacoustic effect that could be attributable to all types of sonic material, and indeed his focus upon *ostinato* figures in particular showed him that such effects are easily transferable from spoken pieces to the idioms of rock music and minimal composition. Regardless, the below observations stem most directly from Johnson's experimentation with speech, particularly his work *John Somebody*:

> Looping, by creating regularity and repetition, pushes the ear towards imposing order on what it's hearing, simplifying the slightly or grossly microtonal variations in speech into the nearest tonal-sounding intervals. It's important to remember that this melodic perception is not an on/off switch—it's a gradation, helped along if the phrase is short, repeated, and the pitches just happen to be close to scaled notes. Given the approximate nature of the pitch and rhythmic content, the perceived tonality and stable beat are only ALMOST there. The brain will fill in the rest of the perception, because it seeks order and comprehensibility.[30]

Johnson's final comment here dovetails neatly into neurological facts brought up by ethnomusicologist Judith Becker in her discussions of musical trance, in particular the claim that the rhythmic nature of synaptic firings causes "different parts of the brain [to be] linked and mnemonic associations [to] develop."[31] Johnson's innovation, having realized our inherent need to seek a rhythmic order that copies that of our biological rhythms, is to augment the repetitive spoken phrases with instrumental parts that cleave close to the perceived 'melodic' structure of those phrases:

> Instruments can provide a framing device which helps the brain along in its natural perception of order—and translation into a simpler pitch relationship than is actually there. It's the power of suggestion, amplifying the natural tendency to organize repeated, limited speech fragments into recognizable melodies.[32]

---

[28] Ubuweb: Sound, Bruce Nauman, http://www.ubu.com/sound/nauman.html. Accessed November 23, 2018.
[29] Ibid.
[30] Suzuki, "A Polypoetical Collision."
[31] Becker, "Music and Trance."
[32] Ibid.

Though such realizations had been made in the realm of experimental media before being considered for academic study, it is nevertheless the landmark psychoacoustic studies of Diana Deutsch that make her perhaps one of the most qualified individuals to comment upon this material. Her 2003 double CD of demonstrative examples, *Phantom Words and Other Curiosities*, focuses singularly upon the speech-to-song illusion that Johnson discusses here, and this has been just one component of her comprehensive mastery of sonic illusions. She has put forward numerous other insights about the relationship of temporal perception to musical perception, which are not necessarily associated with intense repetition but certainly worth being considered in tandem with it. In particular, she has noted the role that musical pitch can play in getting us to understand the passing of time (i.e., "temporal relationships are more accurately perceived between successions of tones that are proximal in pitch than those whose pitches are spaced further apart").[33] Her valuable research will be returned to again later on.

This phantom musicality of repetitive phrasing, which does not have any built-in melodic content, may also be heightened by repetitive material that involves some permutation or re-ordering of a still limited number of phonetic constructions. The multimedia artist Brion Gysin,[34] who inspired Amirkhanian among numerous others, was one of the more incisive individuals to put permutation to use, and creating such effects was an outgrowth of his stated interest in developing means with which to overcome a sort of lexical tyranny over human affairs (see, e.g., his protest that "language is an abominable misunderstanding which makes up a part of matter").[35] Though some allies within the poetic field—for example, Henri Chopin[36]—may not have stated this in such harsh terms, they did remain in agreement when preferring "the chaos and disorder which each of us would strive to master, in terms of his own ingeniousness" to "the Word which everybody uses indiscriminately, always for the benefit of a capitol, of a church, of a socialism etc …"[37]

While later poets in the text-sound idiom would use the insistent repetition of static phrases or phonic values to achieve a malleability of consciousness, Gysin's stock in trade was the 'permutation poem,' in which repetition remained the key element, yet the phrases chosen would have their word order jumbled to, again, 'free' them in a manner that betrayed much about Gysin's direct connections to the Surrealist movement and its equating the freedom of language with the greater emancipation of

---

[33] Deutsch, "The Tritone Paradox."
[34] Brion Gysin (1916–1986) was a painter and poet whose co-development of the 'cut-up' method with author William S. Burroughs was intended to bring writing to the technical level of visual art. This, and other innovations such as the "dreamachine" kinetic sculpture, aimed at profoundly transforming culture at a psychic level.
[35] Ubuweb: Sound, Brion Gysin, http://www.ubu.com/sound/gysin.html. Accessed November 21, 2018.
[36] Henri Chopin (1922–2008) was one of the main exponents of *poésie sonore* or sound poetry, a discipline that he expanded through the use of multi-tracked tape recordings; his *Revue Ou* journal (1964–1972) was also instrumental in the establishment of sound poetry as a medium.
[37] Quoted in Fernandes and Minarelli, *Polypoetry 30 Years 1987 2017*.

mankind. Maurice Blanchot's[38] valorization of Surrealistic language experimentation, for example, had much in common with Gysin's proclamation that poets had come to "free the words": "words are free, and perhaps they can free us, one has only to follow them, to abandon oneself to them, to place all the resources of invention and memory at their service."[39] Blanchot may not have anticipated the degree to which this would happen by vocalization, and recordings thereof, rather than through purely written text.

Gysin's signature pieces have included *Pistol Poem*, in which the multi-dimensional artist recites numerical sequences atop slight variations in the sound of gunfire, and *Kick That Habit, Man* in which permutations such as 'kick that man habit' and 'habit, kick that man' provide an austere yet humorous meta-commentary on repetitive behaviors and the degree of control that one exerts over them, or vice versa (the musicality of these poems has led to collaborations with instrumentalists such as Steve Lacy and Ramuntcho Matta). The 'permutation poem' has since developed a life independent of Gysin's contributions, finding a special place in the repertoire of sound artists who tend to absorb a number of different techniques into a larger experimental program of either re-orienting or disorienting listeners in order to force a kind of holistic self-assessment. The percussionist Z'ev,[40] for example, has made an alchemically inspired use of the same techniques.

While much has been written about how these types of exercises in 'cutting up' reality may result in general changes in behavioral patterns, less has been written about their elicitation of perceptual illusions specifically, and the degree of psychoacoustic effectiveness achieved in the process. It can be argued that select pieces of Gysin's do confer some of the illusions that are referred to in academic literature on the subject. This includes the 'phantom word illusion,' describing a situation in which words, intended to have a purely 'tonal' rather than 'semantic' value, generate the impression that words or phrases have been played repeatedly when this is not, in actuality, the case. *Pistol Poem* might also provide an example of the 'continuity effect,' a.k.a. the 'picket-fence effect' described by Miller and Licklider in 1950. This refers to a sonic scene in which the first sound we hear, interrupted by a much louder one, is perceived to be still audible despite having been rudely cut off (according to Scharine and Letowski, "if a wideband noise burst is inserted into the gap, the tone is perceived to be continuous").[41] *Pistol Poem* achieves this masterfully by adding Gysin's sequential counting to the mix, in effect priming us to interpret a feeling of uninterruptedness in the speech portion even as recordings of explosive sound disrupt that speech throughout the duration of the piece. Nor is this technique utilized only once in Gysin's repertoire: his 1962 piece *Come to Free the Words* seems to be a recording of the artist speaking the same

---

[38] Maurice Blanchot (1907–2003) was a French author and literary theorist known for contributions to the formation of post-structuralist philosophy.
[39] Tone, *Yasunao Tone*.
[40] Z'ev (1951–2017), born Stefan Joel Weisser, was initially active as a text-sound poet inspired by hermetic traditions, before focusing more squarely upon innovative use of found percussion, the latter becoming a staple of what would eventually be known as 'industrial' subculture.
[41] Scharine and Letowski, "Auditory Conflicts and Illusions."

words that he is determinedly scratching out onto a blackboard, and in this case Gysin's forceful vocalizations play the same role as the gunshots in *Pistol Poem* while the chalk scratching plays the subordinate role.

From here, it is not too difficult to bridge the gap between the officially recognized avant-garde represented by individuals like Gysin, and the contemporary popular culture represented by throngs of 'home studio' manipulators. The latter, too, has very recently generated its own manifestations of repetitive persistence where sound is paramount, as in the extensive and somewhat intriguing field of YouTube video mantras/environments featuring hours of repetition of short loops (among a plethora of mesmerizing exercises, prime examples could be: *10 Hours of Darth Vader Breathing, Can You Survive 10 Hours of Patrick Star Asking "Who You Calling Pinhead?"* or *Puddi Puddi 10 Hours*. While examples of this type are generally intended as a form of antagonistic humor, separation from the world of high art does not disqualify them from having similar effects to the works outlined above, nor would they be totally disqualified from acting in a mantric sense, that is, becoming a means through which spiritual energy becomes reified as sound. The original words, as esoteric researcher Phil Hine[42] suggests, "may degenerate into a meaningless mush," but "it is the rhythm which whirls the practitioners' brain towards ecstasy—recognizable by the perception that it is no longer you who chants, but that the chant chants itself through you."[43] Given this understanding that any source material may eventually be severed from its indexicality, simplistic YouTube gags can accomplish the same type of evocative effects as a text-sound composition or a mantra received as part of an ascetic discipline.

2.1.2 Persistence

Another traditional strategy for trance induction—and for the generation of sonic phantoms as we understand them here—is what we would choose to call 'persistence.' Clearly a universal phenomenon, many cultures have developed and established patterns of persistence in their rituals and cultural practices, with said rituals and practices being typified by long duration, insistence, physical and mental endurance, perseverance, and tenacity. This special quality is indeed commonly manifested through repetition; such is the case for a large number of shamanic and ritualized practices that use—and necessarily require—instruments and voices through extended repetition over hours or days without pause or respite.

At this point, however, it should be clarified that persistence does not necessarily imply repetition. As a differentiated mechanism, it is likely best characterized by its effects of saturation, habituation, and endurance. These, in turn, typically lead to different degrees of hypnotic immersion; a fertile territory for uncertainties, illusions, and hence sonic phantoms. Already classic examples of this in Western music are

---

[42] Phil Hine is a British esoteric author and practitioner of the 'chaos magick' as inaugurated by Peter Carroll and his 'Illuminates of Thanteros' organization.

[43] Hine, "The Magical Use of the Voice." See also Hine and Carroll, *Condensed Chaos*.

the work of Morton Feldman[44] (particularly his six-hour-long *String Quartet No. 2*); Stockhausen's *Licht*, a *29-hour Cycle of Seven Operas*; or the multi-hour pieces of Phill Niblock[45] and Eliane Radigue.[46] As diverse as the circumstances are surrounding the creative practices of these composers, and indeed the creation of each of their individual works, all of them can be said to evoke or approach a mental state of the kind that Joachim-Ernst Berendt describes:

> One often has the impression that the same tone sequences are constantly being repeated, but during these repetitions barely noticeable changes take place... The musical phrases and movements of minimal music, its circling, correspond in a fascinating way to the mantras of Asian tradition that in the course of meditation, in a way that is hardly noticeable to the meditating person, begin to develop, grow and work according to their own laws.[47]

As the 'access principle' of the digital information age has spread knowledge of these classics well beyond specialist or avant-garde circles, a whole repertoire of music has eventually emerged that specifically explores persistence and the immersion that it holds out as a possibility. This is often done as a means towards providing a kind of vehicle to enhanced states of consciousness, or at least a vague yet discernible 'presence,' something similar to what Frank Stella[48] suggested as a reaction of critics to minimalist painting: "if you pin them down, they always assert that something is present besides the paint on the canvas."[49] Analogies can be drawn easily enough to sound works in which there is something seemingly present beyond the source material, and in which that source material eventually recedes into the background once a phantasmatic image is formed within the listener's headspace.

An artist such as Sachiko M. (née Sachiko Matsubara), who literally and figuratively made waves with her '100% memory-free' AKAI S20 sampler performing unadorned sine tones, demonstrates this in an unequivocal way: the radically reductionist quality of her music has been likened to a refusal of Japanese collector culture with

---

[44] Morton Feldman (1926–1987) was, along with New York-based contemporaries such as John Cage and Earle Brown, one of the main figures involved with the twentieth-century exploration of indeterminacy in composition, known also for pieces of extended duration, i.e., his *Quartet No. 2* for strings.

[45] Phill Niblock (1933–) is a composer, filmmaker, and photographer who is considered to be one of the pioneers of American minimalist music. Since the 1970s, he has operated the Experimental Intermedia organization in New York City, while also doing much to forward the 'drone' aesthetic in music.

[46] Eliane Radigue (1932–), composer and one-time student of Pierre Schaeffer, is a key proponent of long-form, 'droning,' immersive synthesizer and tape compositions, many inspired by Tibetan Buddhist precepts.

[47] Berendt, *Nada Brahma: The World is Sound*.

[48] Frank Stella (1936–) is an artist working in the media of painting, printmaking, and sculpture, and one of the more widely recognized contributors to the minimalist and abstract expressionist schools of art.

[49] Quoted in Youngblood, *Expanded Cinema*.

its "enumeration of names, trends, dates, and discographical references,"[50] though its true value lies in what it fills this referential void with—namely, with a multitude of psychoacoustic effects that make good upon the promise of Alvin Lucier[51] to 'elicit information' rather than provide it: a pledge that certainly brings to mind again the concept of the 'beholder's share' in aesthetic experiences. With the simplest of tonal means and with minimal physical exertion (the high-register tones used by her require much less sound pressure than low frequency tones would need to be perceived as 'loud sounds'), Sachiko M. interrogates a whole host of innate and culturally received listening conventions, for example providing an illusory sense of immediacy and confounding the natural preference to receive sound from its source rather than the environment.

Her musical approach could be referred to by the classifier 'drone,' which has become taxonomical shorthand for a wide array of instrumental techniques and philosophical approaches animated by the concept of persistence (like 'noise,' the term is not necessarily a pejorative one, depending on the company one keeps). As with the text-sound examples already mentioned, by the eventual perception of movement within structures initially appearing to be static. In some cases, the opposite occurs: music educator Frederick Geiersbach has written about minimalist compositions in which "the V-I cadence ceases to feel final when it becomes the sole ingredient of a piece."[52] Geiersbach is not at all alone in this assumption, and an expansion upon this idea, by David Rosenboom, is worth relating here in full:

> ... in Western music it is generally assumed that a chord, such as the dominant seventh, involves a dissonance that must resolve. However, if one listens to this chord as a continuous drone for a very, very long time, one may find that this unstable structure loses all of its urgency for resolution. It may come to produce a perfectly settled feeling, just as the tonic triad does, particularly if it is tuned so that the frequencies of its individual notes are separated from each other by intervals that approximate whole number ratios (3/2, 4/3, 6/5, etc.).[53]

Given Rosenboom's subsequent comments on how perception of tonal harmony often depends upon "a fixed frame of reference ... not unlike that which prevented physicists from understanding the relativity of the cosmos for so long,"[54] it seems only logical that those who have broken this frame of reference would also dispense with the need to present such music to a proscribed audience. In this respect, the aforementioned habit of injecting the 'avant' tendencies towards persistence with a

---

[50] Loubet and Couroux, "Laptop Performers, Compact Disc Designers, and No-Beat Techno Artists in Japan."
[51] Alvin Lucier (1931–) is an American experimental composer and member of the Sonic Arts Union. His signature works, such as *I am Sitting in A Room* and *Music on a Long Thin Wire*, rigorously investigate psychoacoustic phenomena and the relation of sound to space in particular.
[52] Geiersbach, "Making the Most of Minimalism in Music."
[53] Rosenboom, "Propositional Music."
[54] Ibid.

more streetwise attitude has caused electric guitar-focused bands such as Earth (which, amusingly, was an early working moniker of Black Sabbath) to become cult favorites within indie music circles. Their participation in an established lineage beyond their own genre boundaries seems to make good on previous endorsement of rock music's potential by the titans of the more academically recognized avant-garde (Iannis Xenakis once rhapsodized that "[if the] electric guitars are loud enough, you can hear all the sounds in the universe").[55]

On their *Earth 2: Special Low Frequency Version* LP, the group in question effectively introduced an underground 'rock'-oriented audience to the concept of undiluted, epically sustained tones, which "spread horizontally rather than building vertically, taking on a character not unlike La Monte Young's protracted pieces."[56] Groups like Japan's Boris would eventually follow suit with sprawling guitar drone epics that reached the hour-long mark, all the while making no attempts to disguise their provenance (a 2001 expanded re-release of Boris's *Absolutego* album from 1996 also bears the *Special Low Frequency Version* subtitle). Given that all of these acts aim for an overwhelming sonic output, it is a bonus that duration has a proven psychoacoustic effect on perception of loudness, particularly in sonic material where more than one critical band appears (bands of frequencies in which a second tone will interfere with perception of the first): perceived loudness can change dramatically independent of the total acoustic intensity.

The backhanded critical compliment often directed at Earth and their epigones (i.e., that they "expend more effort redefining notions of time and space than...working out anything as trivial as songs or chords")[57] has become an apt description of many other acts to follow in the band's wake. Sunn 0))), easily one of the most successful and visible of this bumper crop of bands, half-jokingly described itself as an Earth 'tribute band' while dressing up their forebears' sonic motifs in heavy metal theatrics and a self-referential occultist sensibility (complete with druidic robes as a stage accoutrement) that accorded their solid drones a transcendent or alchemical power. The occasional self-identification of Sunn 0))) as 'psychedelic' music is not inaccurate, as much as the cosmetic aspects and packaging of that style may jar with the black-clad, 'left-hand path' occult machinations of so much extreme metal music. If seen as an attitude towards listening rather than a set of genre conventions, 'psychedelicism' is an apt enough descriptor for a style of listening in which listeners' immersion in an altered state of consciousness will allow them to become 'composers' themselves, creating their own personal narratives out of phantasmatic sound.

An intriguing coda to this story can be added courtesy of the Austin, Texas ensemble Waco Girls, who previously staged a complete performance of Satie's *Vexations* entirely with rock instrumentation (and, again, druidic robes), with the vocal accompaniment being a sort of 'liturgical' reading from Dan Nelson's 2008 book of *All Known Metal Bands*. Alex Keller, one of the Waco Girls' constant members, explains the appeal

---

[55] Turner, "Xenakis in America."
[56] Trouser Press, Earth, http://trouserpress.com/entry.php?a=earth. Accessed November 22, 2018.
[57] Ibid.

of merging high volume with intense duration in terms of its perceptually illusory qualities:

> There's no real way to identify the duration in that, because we measure duration, when we're experiencing it over time, in terms of density of events. So a film that has two or three short bursts, it definitely will feel long when you're waiting for the next thing to happen, but in a lot of ways will feel shorter in your overall perception of what occurred, than something else... it might be much more dense. And the idea of listening to something like a drone is like that too, it definitely takes you out of that regular sense of time and puts you into not being able to say "yes, that was a long piece" or "yes, that was a short piece"... especially if it's nice and loud.[58]

In light of what we have already discussed about the phenomenon of auditory streams, what Keller has to say next is illuminating:

> The idea of not having a horizontal chain of events to pay attention to... not having that be a factor that your brain is trying to wrap itself around, means that your brain is going to work harder to find that relationship, you're not listening for a rhythm, you're not listening for events to occur, because events don't occur. If you have a pitch that ascends slowly, over the span of ten minutes, from 1000 hz to 1,100 hz, you can't tell, because you're not given any context, you're not given any reference pitch. That's the way our brains work; we're listening for difference, we're listening for change. If we're trying to listen to change over time and *no change* happens, then all we can do is listen to what's happening in front of us; it's much more immediate, more of a meditative state.[59]

As an aside, some theorists have criticized this aesthetic approach for the reason that it actually fatigues listeners where the intent would be to energize and/or inspire them. While it is certainly not the point of this text to argue in favor of one condition or the other being universal, it is worth introducing this contrasting opinion. Abraham Moles's[60] influential work on information theory is one of the more notable sources in this regard, and follows, as most information theorists would now agree, from the assumption that the formation of 'meaning' must include a degree of redundancy (see, for example, Donald MacKay's famous suggestion "a distinction that makes a difference," which was later adapted by Gregory Bateson into the more memorable "a difference that makes a difference"). It would then stand to reason that a process in which 'events don't occur,' at least on a perceptible level, is equal to one in which no redundancy exists (i.e., each moment spent attending to the piece/work in question is equal to the next and preceding moments in communicative potential). Moles's

---

[58] Personal interview with the author Thomas B. W. Bailey, November 30, 2018.
[59] Ibid.
[60] Abraham Moles (1920–1992) was an acoustic/electrical engineer and information theorist, pivotal in that field for compiling an extensive bibliography that investigated information theory from an aesthetic point of view.

somewhat harsh assessment of non-redundant works as "lacking interest" because of their "uniform probability distribution for all [their] elements" is likely grounded in the assumption that this erases their communicative value, since even the most minuscule unit of meaningful data comes to be defined by its lack of uniformity.[61]

However, Moles's criticism can be countered in a number of different ways. One might be the fact that extreme complexity, present in musical genres ranging from 'hard bop' jazz improvisation to apparently unstructured noise, also presents some listeners with a kind of fatigue or eventual 'lack of interest,' a fact that cannot be explained by lack of redundancy. It is also clear that the lack of redundancy in question is not dissimilar from what test subjects in Ganzfeld ['total field'] experiments[62] report; this being an experiment in Gestalt theories that provides a visual experience of completely uniform light in which there is no incoming temporal and spatial information. During this time "the only significant perception" may be "the perceptual process itself"[63] or the visualization of entoptic phenomena[64] (that is to say, even in this situation there is still no blank 'void' of perception).

In short, Moles's criticism also seems not to account for the power of the phantasmatic; for ritualized and fully immersive experiences to be intense enough that illusory percepts make up for the initial lack of meaning. We might say that such an experience occurs because of the introduction of *aesthetic* content, which will reveal *semantic* content over time: an experience that is often entered into without the expectation that the latter type of content will not be understood immediately or as quickly as it might be in other types of experiences. William Smith, speaking on a different type of non-redundant artwork, lays this out as follows:

> Rather than a standardized repertoire of words and sounds, individual receptors (people) bring to a work of art their own unique set of experiences. These experiences, rather than an abstract set of rules, configure different aspects of an artwork as redundant or original. To use an example that Moles cites, someone who has heard Beethoven's Ninth Sonata multiple times would have a different, presumably more subtle, schema for making a distinction between originality and redundancy than someone who hears it for the first time, which is not to say a work can ever be fully depleted or made totally redundant, after multiple repetitions. Instead, the Beethoven connoisseur can draw from the work information that will be different from what will be communicated to the novice.[65]

---

[61] Moles, *Information Theory and Esthetic Perception*.
[62] This is a technique used in parapsychology (and neurophysiology) to test individuals for extrasensory perception. This experiment can be carried out by simply placing the two halves of a dissected ping-pong ball over test subjects' eyes. See Carpenter, "ESP Findings Send Controversial Message Siri Carpenter."
[63] Daly, "The Ganzfeld as a Canvas for Neurophysiologically Based Artworks."
[64] Entoptic phenomena are visual effects caused by the eyes own structures, for example eye floaters.
[65] Smith, "A Concrete Experience of Nothing."

In this context, the epitome of duration is, of course, the work of American composer La Monte Young, whose aesthetic influence can be seen in other notable examples of the persistence aesthetic such as Alan Lamb[66] (Lamb's work has followed from Young's fascination with the sound of 'singing wires,' which in Lamb's own work act to excite instruments such as gongs or flutes, which are excited by the wires in an aleatory fashion). Young came to prominence via projects such as his ensemble The Theatre of Eternal Music and his ongoing composition/immersive environment of *The Dream House* in New York City, in which electronic music plays throughout the entire day across the six floors, and the interiors are lit with the magenta glow of his partner Marian Zazeela's light projections (this installation/environment has been continuously unfolding for over 25 years now). The many discussions on the topic of 'infinite' duration, spurred on in part by Young's work, have provided a much-needed sense of continuity between ancient 'Eastern' traditions animating modern 'Western' ones, particularly the way in which the former has utilized sustained drone in order to enable extremely complex modal scales along with an accompanying variegation of subtleties in mood. More so than formal considerations, though, Young's true legacy may lie in his helping to establish what Pauline Oliveros[67] described as 'deep listening' in opposition to modern visuo-centrism (e.g., in his *Composition 1960, #5*, whose sole 'performer' was a butterfly loosed in the audience, and which was explicitly meant to encourage listening to objects and events typically apprehended with the visual sense only).

It can be argued that this sort of heightened reverence accorded to sound, that is, the refusal to understand it as being secondary to some other form of sensory information, is one of the impulses that most attracts and influences individuals to the modern revival of a 'persistence' aesthetic in composition. Both the *Dream House* and *Theater of Eternal Music* projects, for example, have attempted something along the lines of what Ligeti[68] described as the impetus for his composing the piece *Lux Aeterna*:

> the words [lux aeterna, or eternal light] only served as a chance for me to compose music which is in fact musica aeterna: as if it has been sounding from time immemorial and would be going on for ever—we only hear a part of it. It emerges from nowhere, it is here and slowly disappears.[69]

With this in mind, Young conceives of his entire creative output as a single 'timeless' work, with one significant slice of that—his *Well-Tuned Piano* piece—exploring the poetics of time in such a way that listeners would need to listen for a full lifetime

---

[66] Alan Lamb (1944–) is a Scottish-born, Australian composer specializing in coaxing resonance out of 'singing wires' by a variety of different methods.
[67] Pauline Oliveros (1932–2016), a composer whose 'deep listening' approach encouraged a greater consideration of environmental phenomena during performance and recording, also acted as the director of the San Francisco Tape Music Center.
[68] György Ligeti (1923–2006) was an Austrian–Hungarian innovative and influential composer essential to the development of 'micropolyphony,' an approach towards musical voicing that favored vertical tone clusters.
[69] Quoted in liner notes from *Atmospheres* [CD], György Ligeti, Berlin: Deutsche Grammophon, 2006.

in order to hear two repetitions of the same musical interval. Certain sounds when experienced in a sustained form allow for a greater spatialization of these sounds, not only reintroducing us to the physicality of them but also (in Young's reckoning) doing this in such a way that the sounds can be said to have a clearly defined structure, that is, with their own distinct exterior and interior surfaces. Composers such as Amacher have described this feeling of heightened attentiveness as 'perceptual geography,' a concept she applied to her work with 'additional tones' in general, and otoacoustic emissions[70] in particular. Young, meanwhile, describes this experience of 'getting inside' sounds as follows, touching upon other perceptive realities such as synesthesia or cross-modal perception in the process:

> Sometimes when I was making a long sound, I began to notice I was looking at the dancers and the room *from the sound* [italics ours] instead of hearing the sound from some position in the room. I began to feel the parts and motions of the sound more, and I began to see how each sound was its own world and that this world was only similar to our world in that we experienced it through our own bodies, that is, in our own terms. I could see that sounds and all the other things in the world were just as important as human beings and that if we could to some degree give ourselves up to them, the sounds and other things that is, we enjoyed the possibility of learning something new. By giving ourselves up to them, I mean getting inside of them to some extent so that we can experience another world. This is not so easily explained but more easily experienced.[71]

The experience described here has been best demonstrated, again, via the *Theater of Eternal Music*: it was a concept that found a particularly adept ambassador in the figure of Tony Conrad,[72] whose participation in it would eventually inform his work in other media and earn him recognition as one of the key shapers of American minimalism and adjacent phenomena such as structural film. The meditative or yogic discipline required of this ensemble, which required all the assembled players to play in just intonation for extended periods of time often resulted in a phantasmatic effect of the type that affirmed Young's above comments on 'getting inside' of sounds. Conrad described this phenomenon in detail to his biographer Brandon Joseph:

> I found that just when I was playing in tune, I would listen again, and I would hear even more minuscule artifacts in the tone that would be moving around. Maybe they're beats between the harmonics, for example, and you know that way up the harmonic series, if you have a little inaccuracy, you're gonna be generating

---

[70] Otoacoustic emissions, first demonstrated by David Kemp in 1978, are a form of sound generated by the inner ear; effectively making the ear a transmitter as well as a receiver. This concept, and its particular relation to the phantasmatic, will be discussed in greater detail in this book's closing chapter.

[71] Young and Zazeela, *Selected Writings*.

[72] Tony Conrad (1940–2016) was a New York-based experimental multimedia artist whose activities included many of the innovations in structural film and minimalist music of the 1960s.

beat tones that are quite off. So the more you get closer and closer, the more you begin to hear things lining up and then things not lining up. And a kind of hearing became necessary which I can only describe as going into the sound, going into the interstices of listening more carefully in the way that you can learn to discriminate like when you hear that there's a number of pitches in one note...[73]

Conrad seems to have gone to some effort to differentiate himself from public perceptions of Young's spirituality—the latter being exemplified by pronouncements such as "universal truths are being transmitted directly through me,"[74] while at the same time unintentionally differentiating himself from the Tibetan Buddhist orientation of players such as Eliane Radigue and Pauline Oliveros (Oliveros's overtone-driven *Tashi Gomang* falls into this category, as do numerous long-form works of Radigue). Namely, Conrad felt the effects of the *Theater of Eternal Music* were not due solely to "the mythical rigor of Eastern mysticism" but rather to a more non-denominational, inter-disciplinary "excitement of a continuing growth."[75] When the theoretical and spiritual scaffolding falls away, however, what remains is an impression that all the above-named players have contributed to a single raging current of phantasmatic sound, one that not only connects a non-capricious engagement with modernity to more firmly rooted traditions, but also connects human communications to that of the natural world. This will be discussed more in the section that focuses more squarely on the latter, but for now it may suffice to note that the effects Conrad mentions above are commonly found in the 'choral' behavior of amphibian and insect species across the planet. Field recordist Bernie Krause describes the effects of intermodulation as follows:

> ... when two or more signals are so close in pitch that they occasionally beat against each other, momentarily cancelling each other's signal, a totally different acoustic effect than any of the original sources sounding individually.[76]

By now, enough artists have taken the lessons of persistence to heart, and ignored the ironically persistent criticisms of their own persistence aesthetic (e.g., not bringing enough 'original' or 'non-redundant' information to the table), to form a cultural movement spanning several generations of musicians. German 'kosmiche' bands such as Popol Vuh[77] and Tangerine Dream,[78] as well as British artist Brian

---

[73] Joseph, *Beyond the Dream Syndicate*.
[74] Quoted in Chave, "Revaluing Minimalism."
[75] Joseph, *Beyond the Dream Syndicate*.
[76] Krause and Books, *The Great Animal Orchestra*.
[77] Popol Vuh, named after the Mesoamerican mythological text with that title, were the musical project of Florian Fricke and one of the more successful artists in the German "krautrock" canon of mind-expanding, electronically enhanced rock music.
[78] Tangerine Dream, another key exponent of the "krautrock" school, were led by Edgar Froese and are credited with being key ambassadors of synthesizer-based composition into the pop realm, as well as incorporating that same style into New Age ambience.

Eno[79]—and before them some of the abovementioned artists (Niblock, Radigue, Young)—are considered to be at the roots of this genre's genealogy. Today this would include the work of hundreds of artists worldwide (at the very least, enough to be cataloged on omnibus recordings such as the *Swarm of Drones/Storm of Drones/Throne of Drones* trilogy from the Asphodel label, which is already something of a 'classic' of the genre). The increased number of artists working with this kind of prolonged sonic materials has recently been supplemented by gradual stretching of the form's temporal dimensions and ever-deeper dives into the heart of its immersive qualities.

In this context, a recent prime example of persistent immersion without repetition is the piece *untitled #305 [seven nights]* by composer and audio-artist Francisco López, a fifty-six-hour long drone piece released on an SD memory card. This is no mere footnote to the story at hand: after all, an example such as this illuminates the degree to which digital technology paradoxically enables an immersive form of listening that would have previously been confined to on-location listening in the natural world. The ability to release or publish days' worth of audio compositional material on a tiny storage device with a capacity in the gigabytes of information has become something of a hallmark for López (*untitled #305* on the Somnimage label was preceded by *untitled #272*, a 24-hour piece based on location recordings from Antarctica). Advancements in non-volatile memory have enabled an emergent movement concerned with broadening the temporal possibilities for immersive listening. As the idea of partitioning slices of the sonic experience via 'albums' becomes more redundant, it is very likely that more artists will take full advantage of intensely long-form composition and, in turn, commit even further to an aesthetic of persistence.

In what concerns the pieces I present in this volume as part of the *Sonic Phantoms* ongoing series of compositions, I have used persistence to a greater or lesser extent in all of them, but it has been a particularly relevant strategy in my *Harp Phantoms*, *Drawing Phantoms*, and *Vocal Phantoms* compositional series. In these cases, I have explored and promoted ritualization and trance-induction through—among other things—extended duration.[80] An extended temporal framework is essential to promote the necessary engagement and perceptual openness to give rise to heightened states of sonic experience that, to some extent, bypass or inhibit our usual cognitive-aware states (inner dialogue, critical judgment, etc.). Persistence has been for me a way to entrance both listener *and* performer, to absorb or immerse them simultaneously in a state of embodied participation in the music.

---

[79] Brian Eno (1948–) is a composer and producer to some of the world's most enduring pop musicians (David Bowie, Talking Heads, etc.), also widely credited with bringing the term 'ambient music' into common use.

[80] A recent example is *CyberOpera* by the author on USB memory card, 2019, duration three hours eighteen minutes.

2.1.3 Layering

In my compositional work, there is a kinship to much of the work mentioned above, in that there is marked emphasis on vertical modes of organization as opposed to horizontal forms of sonic structure. By 'vertical' I mean simultaneous layering of sonic patterns, rather than sequential and temporal organizations. Focusing on verticality seems to help to promote a multi-perspectival quality of listening: namely, one has time to 'zoom into' details and shift perceptual focus onto alternate pattern combinations. This strategy provides for the opportunity to 'inhabit' the material, as it were, inducing a shifting of attentional focus between overall sonic appearance and the ever-changing details of features that can be appreciated through this perceptual focusing and defocusing upon patterns rising and submerging within the textural continuum. This multiplicity of perspectives on the same material generates an ideal fertile ground for sonic phantoms to arise.

Many of my works are constructed to form a tapestry of simple, yet multi-layered component elements or patterns. Stratification and simultaneity of such layers can give rise to rich levels of complexity and intricacy. One of the ways in which this can manifest is by what one might describe simply as 'densification'; that is, a significant accumulation, accretion, or build-up of simultaneous sonic layers that give rise to emergent textures and sonic fields. I have employed densification to some degree in many of the pieces associated within the larger *Sonic Phantoms* undertaking, to the point where it becomes the foremost mechanism for the induction of sonic phantoms.

Achieving a satisfactory level of densification can be done by means that, in our current music production realm characterized by instant access to digital audio workstations, seem simple or mundane. One of the most cited examples is the early tape music experimentation conducted by Steve Reich: his tape loops of short yet evocative phrases, when played back on multiple tape recorders going slightly out of phase, produced a startling degree of alteration in the perceived phonic qualities of those same looped phrases (see for example, the classics *It's Gonna Rain* (1965) and *Come Out* (1966)). Such pieces are often cited as examples of how one can make truly aleatory or self-composing music with the most quotidian means, though they are equally valuable for their demonstration of how subtle changes in identical audio components can create an illusion of greater density than what is actually being played or projected (it is not uncommon for listeners to Reich's 'tape loop' pieces, as well as pieces that utilize a similar technique, to hear a 'third' distinct auditory stream in addition to the two actually presented). Trevor Wishart,[81] himself no stranger to such effects within his own compositions, is one of these listeners, and describes the singular effect of this piece as follows:

> In Steve Reich's *Come Out* the phrase 'come out to show them' is used purely as a timbral motif... as this [process of synchronization and gradual de-synchronization]

---

[81] Trevor Wishart (1946–) is a British composer, vocal improvisor, software designer and educator (with printed educational works including *On Sonic Art* and *Audible Matter*).

happens, rhythmic and timbral patterns (partly due to phasing effects) are established, which arise directly out of the timbral properties of the sonic object 'come out to show them' (or rather the specific speech utterance of this phrase initially recorded on tape).[82]

Elsewhere, Dean Suzuki asserts that "Reich's pioneering use of tape heightened the musical qualities of speech—melodic, rhythmic and timbral characteristics—vis-à-vis repetition."[83] As Reich contributed to this developing lineage, he did not limit the application of this technique to taped recordings of the human voice, and used a more or less similar technique to make distinct pieces for violin (*Violin Phase*) and for handclaps (*Clapping Music*). More pertinent, though, is how this process has served to elegantly confirm points made earlier by Johnson and Ter Veldhuis with regard to the illusory mental conversion of tonally 'flat' speech into song structures with a greater variation in pitch.

It did not take long at all for a sort of lineage to develop around pieces such as Reich's and to become more ambitious in their content, while still cleaving close to the concepts of intuitively similar yet out-of-phase phrases. One notable example comes to us via Stefan Weisser (later Z'ev) and the voice transformation pieces that he busied himself with up until the early 1980s. *Book of Love Being Written as They Touched*, from 1975, is indicative of the general approach that Weisser took: a vocal recitation in which eight voices each read sentences beginning with one of the eight words in the title. The permutative nature of the results [8! = 40,320 total permutations] led eventually to a condition of incredibly polyphonic density that, it could be argued, conferred an illusory sense of there being far more voices involved in the recitation, as well as 'additional' phonetic or even semantic content exceeding the simple phrase actually being recited. The composer would later see fit to merge this concept with that of persistence, crafting a three-hour piece *Oomoonoon: Dancing on the Brink of the Word #1*, in which three-minute loops from the three reciting voices would have their ambience recorded and then played back in further permutations.

Within contemporary music there is, of course, a vast repertoire of further compositional examples informed by an explicit and thorough exploration of such a 'layering' sonic territory, from György Ligeti's *Poème Symphonique (for 100 Metronomes, 10 Performers and 1 Conductor)* to Krzysztof Penderecki's[84] *Threnody (for the Victims of Hiroshima) for 52 Strings* to Paul Dolden's[85] massively multi-layered instrumental compositions (notably those encountered in his album *L'Ivresse de la Vitesse*). Ligeti, in particular, is worth returning to as a source of some of the most meaningful works in this creative realm: besides *Poème Symphonique* there is also *Atmosphères* (famously used without the composer's approval in Stanley

---

[82] Wishart, *On Sonic Art*.
[83] Suzuki, "A Polypoetical Collision."
[84] Krzysztof Penderecki (1933–) is one of the preeminent Polish composers of modern music, known particularly for the aforementioned *Threnody*... and his *St. Luke Passion* of the mid-1960s.
[85] Paul Dolden (1956–) is a Canadian electro-acoustic composer, known for innovative deviations from the more 'traditional' tape-based music associated with the genre.

Kubrick's *2001: A Space Odyssey*) and *Continuum*. Naturally, Ligeti's oeuvre has always been a rich territory to mine for phantom effects, especially given the composer's explicit encouragement of "perceptual illusions as musical devices in their own right,"[86] and his confirmation of taking this viewpoint as a deliberate aesthetic strategy. Among other things, this exploration of illusion has included the translation of written/scored polyphony into apparent harmony. *Continuum*, meanwhile, has been introduced by the composer as a piece that contains a sort of 'phantom' sense of rhythmicity following from its pointillistic buildup of isochronous pitches, and from this follows a psychological illusion of apparent motion. It is an illusion that

> gives rise to the perception of streams, i.e. progressions of tones that are perceived as belonging to the same constant sequence... this phenomenon relies on the rate of presentation of individual events and is most clearly illustrated when events are played rapidly, giving rise to a granular texture...[87]

This latter description also brings to mind the work of Iannis Xenakis, in particular *Persepolis* (which, at fifty-six minutes, is the composer's lengthiest work apart from the 1969 *Kraanerg*). This is another such piece that works upon principles of layering, though the same can be said as well for the many contributions to the electro-acoustic canon that formed a significant part of Xenakis's legacy. Described fairly accurately as being "constructed from eleven textures, each developed independently and distributed across the eight channels of the tape,"[88] the resulting perceptual confusion creates a situation whereby "it is not easy to locate the sectional divisions, as different channels shift at different times and the dominance of one sonority over the rest is statistical rather than clear-cut... it is hard to identify the sources of the sounds, too, but they can be distinguished by spectral definition, continuity or discreteness, and register."[89] This is by no means an anomaly where Xenakis' oeuvre is concerned, with similar effects predominating in pieces spaced years apart from one another on the composer's timeline (see, e.g., *Bohor* from 1962 and *La Légende d'Eer* from fifteen years later). Given what has just been said about Ligeti's development of auditory 'streams' arising from an overwhelming saturation of discrete events, it is worth noting Xenakis's own very similar interest in achieving this effect, which dates at least to his initial tape piece *Diamorphoses* (1957): "... by dense mixing one can obtain continuous sounds out of discontinuous ones. It also became clear that there is a logarithmic relationship between the increase in density and its perception."[90]

It is also worth discussing the experimental procedure by which the composer reached this conclusion. In the initial stages of *Diamorphoses*, a tape was recorded that featured sounds of an 'irregular density,' which were then duplicated and mixed together with the original tape. The duplicate tape was edited so as to delete any

---

[86] Pressnitzer, Suied, and Shamma, "Auditory Scene Analysis."
[87] Cambouropoulos and Tsougras, "Auditory Streams in Ligeti's Continuum."
[88] Harley, "Book Review: Persepolis."
[89] Ibid.
[90] Xenakis and Varga, *Conversations with Iannis Xenakis*.

instances of echo or repetition, and the resultant multi-tape composition auditioned by the composer to see if he could spot the difference between the two sources. This experiment was repeated again with a set of three tapes, and with still more, though the composer noted that, no matter how perceptibly great the increases in density, there would only be one perceptible step: the "logarithmic connection" Xenakis mentions was confirmed when noticing that an increase in density that should have been perceived as "nine times greater" was only audible as a single step up from the previous iteration of the same composition.

The other major manifestation of sonic layering is 'interlocking,' an aesthetic and psychoacoustic device that again manages to impart archaic knowledge to technological implementation. The interlocking of simple elements is of key compositional interest for me in my own quest to produce sonic phantoms and is at the core of all the compositions in my *Instrumental Phantoms*, as well as *Vocal Phantoms* and *Natural Phantoms* compositional series (described in the corresponding chapters below).

The general attitude and aims related to interlocking have, like the phenomena of persistence and repetition, animated entire cultural curriculae—not necessarily being limited to audio material. For example, in modern times, this technique has been illustrated by the rapid interchange of 'on-off' states presented by the films of Conrad and Sharits,[91] and the viewing of Gysin's 'kinetic sculpture,' the *Dream Machine*. The enduring appeal of stroboscopic lighting within concert and dance club scenarios also testifies to a fascination with viewing the alternation between 'something and nothing,' and with either discovering or rediscovering the ways in which our senses manage to compensate for lack of either aesthetic or semantic information. A natural progression has followed in the type of audiovisual synthesis popularized by artists such as Ryoji Ikeda,[92] Carsten Nicolai,[93] and the formative roster of the Raster/Noton record label, in which two-dimensional imagery and motion graphics reinforce the messages already being conveyed by the aforementioned rapid exchanges in the musical content. Several key albums released by Raster/Noton at the dawn of the 2000s featured methodically, even clinically precise latticeworks of sonic presence and non-presence delivered in a number of ways, that is, identical blocks of tone alternating rapidly between left and right speakers, or equally non-gradual, binary switching between a near-ultrasonic tone and another tone bordering on the infrasonic.

Paul Sharits, in explaining why he deliberately employed an equivalent sort of effect, stated that "where the visual image is redundant, the auditory image is active, and as the visual image becomes active (begins falling), the auditory becomes redundant."[94]

---

[91] Paul Sharits (1943–1993) was an American filmmaker identified, along with contemporaries such as Michael Snow and Tony Conrad, with the genesis of the 'structural film' movement, itself a precursor to later forms of video installation art.

[92] Ryoji Ikeda (1966–) is a Japanese sound artist, and participant in the performance group Dumb Type, whose audio work focuses upon extremes of audible frequency and their psychoacoustic effects.

[93] Carsten Nicolai (1965–) is a sound and installation artist, and co-founder of the Raster/Noton media publisher, occasionally using the alias Alva Noto.

[94] Quoted in Cornwell, "Paul Sharits."

The same logic that Sharits applies to the alternating of sensory modes also applies to changes perceived within a single mode, in this case interlocking patterns which are alternating patterns of sound and silence between several instruments; voices or sounds that combine to make a 'whole' or emergent Gestalt. They are usually dovetailed and in many cases equally spaced in time, but the paradoxical result is typically an outstanding global complexity. Intricate sonic tapestries emerge from simple repeated patterns—occasionally magnified by fast-paced, machine-like repetitions at the limits of human perceptual temporal resolution—that form larger interlocking and complex structures. A prime example of the accelerated, densely evocative and hypnotic structures described above is that of the traditional *amadinda* and *akadinda* xylophone music of the Royal Courts in Uganda. The essence of this kind of pattern construction is that no one part contains all of the notes or sonic elements, but instead they share common elements, such as common borders.

This last point returns us to the 'figure and ground' relationship already discussed: if such music strikes the listener as compelling, it may have something to do with the phenomenon as Ramachandran and Hirstein describe it: "information (in the Shannon sense) exists mainly in regions of change—i.e. edges—and it makes sense that such regions would, therefore, be more attention grabbing—more 'interesting'—than homogeneous areas."[95] The aforementioned 'common borders' are as captivating to us, in a neuroaesthetic sense, as the visible edges that act as our guides in discerning discreet patterns. Sonic phantom patterns produced by interlocking are always a resulting image of the perceptual processes of grouping or segregation of the component parts.

From a certain perspective, *sensu lato*, most music obviously contains some degree of interlocking. It is quite rare that a musical listening experience involves hearing isolated, pure tones; even in the age of music synthesizers, which enable such an experience far better than earlier pitched instruments, nearly all music that we encounter features two or more frequency components. Distinctive cases of sonic phenomena such as sonic phantoms, acoustic illusions, or explicit aural ambiguity, however, will take center stage when interlocking constitutes the core of the compositional process and generates an internal structural world; when it is the focus of music-making to the point of becoming the aesthetic and structural essence of the piece created. Maryanne Amacher's compositional approach provides one definite affirmation of this aesthetic: in her reckoning, it is the process of 'interplay' that creates what is to be the true focus of any resultant piece, namely an ability to "'know' what acoustic intervals are indeed affecting responses in our ears and brain," and also to listen not necessarily to the primary musical material but to those processes that are "innate and perhaps distinctly human capabilities."[96] Curiously, this conceptualization of music as a vehicle to initiate more far-reaching and enduring perceptual experiences returns us to the old theriomorphic or metaphorical descriptions of the shaman's musical efforts as a 'horse,' 'canoe,' and so on, and shows a remarkable continuity between the aims of archaic traditions and those of modern experimental music.

---

[95] Ramachandran and Hirstein, "The Science of Art."
[96] Amacher, "Psychoacoustic Phenomena in Musical Composition."

A general analysis of different types of instrumental and vocal music from around the world (both traditional and non-traditional) in which we find examples of prominent sonic phantoms reveals that these are often either a by-product of, or in many cases intentionally produced by, interlocking techniques such as hocketing, to which we will now turn. There is some argument over the etymology of this term, which, in a roundabout way, actually helps to provide a clearer conception of this vocal technique for those who are not already familiar with it. The term is traceable alternately to the Arabic *al-quat'* (where it refers to a 'cutting' process) or Latin *hoquetus*, and becomes more vague still in its French adaptations: *hoquet* can mean a 'hiccup' or 'stutter,' though the more likely candidate for a musicological translation seems to be the verb *hochier* ('to shake'), which shares orthographic kinship with similar terms in Middle High German and Middle Dutch. The 'shaking,' in this sense, refers not to some sense of instability or an involuntary shudder, but to the vivacity involved in the act of hocketing; a term perhaps more in keeping with modern developments in audio production might be 'oscillation.' Keeping this in mind and fusing it to all the above definitions, we have a rough outline of a vocal technique defined by 'choppy,' rapidly alternating rhythms. Musicologist William Dalglish, who impassionedly defends the technique against claims of being "a technique of purely subordinate import" and laments how the hocket has "never [been] precisely defined by modern authors,"[97] essentially confirms this with his own definition of the hocket as "the truncation of a melodic line with rests and the distribution of its tones between two or more alternating parts."[98]

The historical origins of the hocket technique are as open to debate as the origin of the term itself: the technique is generally agreed to have originated in antiquity,[99] commonly found in traditional musical cultures around the world as in the flute and trumpet ensembles, xylophones, drums, whistles and horns of sub-Saharan Africa, flute dance music of New Guinea, panpipe playing of the Andes and the polyphonic singing of the Swiss Alps and the Georgian Caucasus. Within the specific confines of the Western tradition, the hocket arguably realized its full potential in the polyphonic chant of the Middle Ages. It was typically integrated into forms such as the *organum*, *conductus*, or *motet*, while flourishing in the Notre Dame school and in the hands of composers such as Perotinus,[100] though it did in time become branched-off into stand-alone pieces also called 'hockets.' The technique was situated with a larger trend of what Dalglish identified as "flamboyant and unorthodox singing in church,"[101] and, more interestingly, formed part of a not unanimously respected tradition of "amazing misshapen shapeliness and shapely misshapeness."[102] This is as good a

---

[97] Dalglish, "The Hocket in Medieval Polyphony."
[98] Ibid.
[99] Ibid.
[100] Perotinus Magnus was a late twelfth-century composer of the Notre Dame school of polyphony, and author of the *Magnus liber organi* [Great Book of Organum], *c.* 1160.
[101] Dalglish, "The Origin of the Hocket;" and see Dalglish, "The Hocket in Medieval Polyphony."
[102] "this shapely misshapenness, this misshapen shapeliness?" attributed to Bernard of Clairvaux in the *Apologia ad Guillelmum* written in 1125.

description of musical ambiguity as any, yet also indicative of how these techniques rankled contemporaneous thinkers such as St. Bernard of Clairvaux, occasioned intense polemics against medieval singing in general, and even inspired at least one papal bull (Pope John XXII's *Docta Sanctorum Patrum* of 1324). At this time, the form could be subdivided into *duplex*, *triplex*, *quadraplex*, and *contraduplex*, which referred respectively to one part alternating with another, two parts alternating with a third, four parts in a free exchange, and two alternating pairs. The latter two types were, throughout the thirteenth and early fourteenth centuries, typified by intense speeds, and it might be speculated that this increased acceleration eventually led to a kind of burnout or fatigue with the form, as the isorhythmic approach was conspicuously absent from polyphonic music by the fifteenth century.

However, as mentioned above, the hocket is not to be bound to a single geographic region and historical period. The aforementioned cultures embracing the hocket developed the technique independent of one another, and these efforts spanned multiple continents (even after its being phased out of 'official' medieval church music, it continued to be used within various types of roughly contemporaneous European folk music). On the African continent alone, there are several distinct musical cultures and instrumental configurations that make use of it. There are enough, in fact, that an exhaustive inventory would derail this discussion, yet all the same there are a few key examples emblematic of the common effects that this style aims for. Hocketing features noticeably in the music of the Kpelle people from Guinea/Liberia/Ivory Coast (e.g., horn ensembles, bush-clearing work songs, funeral songs), the Aka (Babenzele) from Central African Republic and the northern Republic of the Congo, Baka people from Cameroon and Gabon, the *mbira* music of the Shona of Zimbabwe, *nyanga* panpipes of the Nyungwe people in Mozambique, Banda-Linda Horn ensembles from Central African Republic, and many more besides. Variants on the technique are as geographically distant as the blazingly quick speed of the Kiganda xylophone music, and the wind ensembles (typically involving flutes with a limited tonal range) adopted by Western African peoples such as the Builsa of Ghana. The Western African traditions, in particular, provide a notably different conception of the hocket than that provided by Western liturgical music: according to musicologist J. H. Kwabena Nketia, it is because these musical cultures elevate the hocket from a *device* to a *technique*. That is to say, the hocket as it appears throughout Africa is a functional feature of the music rather than an ornamental flourish, or "a means to an end, not an end in themselves."[103]

Nketia speculated that the hocketing technique is an ingenious method of "fill[ing] in breaks," or providing a sense of continuity between melodic parts: the hocket essentially provides an illusion of a fuller sound where there is an "absence of harmonizing notes" or a "necessity for throwing the melodic line into relief."[104] By doing little more than alternating brief rests with equally brief bursts of sound, or having a succession of single players sound off while the rest of the ensemble remain mute, this effect can achieve dramatic results in terms of locking in a listener's focus. As

---

[103] Nketia, "The Hocket-Technique in African Music."
[104] Ibid.

Nketia affirms, this emphasis upon the resultant aspect of the music, the binding into an 'integrated whole,' is very much in keeping with the phantasmatic trick by which musical effects become an experience unto themselves, a secondary experience arising apart from apprehension of the 'original' experience. This same technique, when used within the Asante *ntahera* trumpet ensembles of Ghana (a kind of heraldic ensemble that provides court music for Ashanti chieftains), provides a clear phantasmatic effect that Nketia identifies as "imitating the falling intonation used in speech...a major function of this instrument is to 'talk' while the others are playing."[105] Although he warns about the limitations inherent in the hocketing style as a whole—especially the fact that such a tightly controlled weave of sound makes improvisation nearly impossible—this illusory effect seems to adequately compensate for whatever sensory revelations might otherwise be denied by this.

This is a concept that brings us to one of the truly interesting manipulations of the hocket, namely its use by both vocal and instrumental performers within the same piece, which resultant mixture reinforces an already profound mesmeric effect provided by interlocking. One of the more unequivocal, immediately recognizable, and uncannily modern representations of this technique is the Balinese *kecak* or 'monkey chant' accompanying the gamelan-led dance drama. *Kecak* has been utilized in cinematic ventures as diverse as Fellini's *Satyricon* and the classic anime film *Akira*, testifying fairly well to its adaptability to a variety of dramatic environments as well as to the far-ranging appeal of interlocking techniques. The onomatopoetic male 'cak' chorus that has featured so prominently in film soundtracks, and which is otherwise known as the *gamelan mulut* or 'mouth gamelan,' is memorably described by sonic researcher David Rothenberg as such:

> ... each voice makes a cricket-like cha sound, all hocketing back and forth in an exciting fast beat, regular and irregular at once—CHA! cha cha CHA cha cha cha CHA cha cha cha cha cha cha cha cha CHA cha cha cha cha CHA cha cha cha cha cha cha ...—while above, a drawn-out droning voice chants tales from the Ramayana, one after another for many hours of ritual intensity.[106]

This vocal technique is regularly reinforced by another form of hocketing instrumentation, the ringing and rippling metallic sonorities of Balinese *gong kebyar* gamelan. Whether the hockets manifest in vocal or instrumental form, they represent what Bateson identified as the key unifying facet of that culture's music and of the 'Balinese character' as a whole: "modifications of intensity determined by the duration and progress of working out...formal relations" rather than the more familiar Occidental structure dependent upon "rising intensity and climax."[107] Running counter to Bateson's thesis is Kendra Stepputat's disclosure that *kecak* is a 'tourist genre' less than a century old, which was intended as a fusion of the tastes of native Balinese

---

[105] Nketia, "The Hocket-Technique in African Music."
[106] Rothenberg, *Bug Music*.
[107] Bateson, *Steps to an Ecology of Mind*.

artists and Western expatriates[108]—yet this certainly does not disqualify it from being a worthwhile object of study, nor from its being part of a considerable historical lineage of hocketing techniques *and* music intended to accompany trance states (the latter is a subject to be covered more in depth soon). With this in mind, Stepputat's recognition of *kecak*'s essential perceptual characteristic—that is, the "complex interlocking structure" occurring during the palpably dense *pola cak* chorus—is more important to note than her dismissal of the form's 'deep' origins.

Other diverse examples include the hocketed and yodeled vocal polyphony known as *Krimanchuli* in Georgia,[109] the Lithuanian *Skudučiai* multipipe flute music, and lesser-known Russian panpipe traditions from the Kursk province.[110] Musical examples such as these, although from very different parts of the world, have much in common in that they all utilize the principle of interlock and give rise to distinct sonic phantoms as a result. It is thus a useful way to generate highly complex results from simple and limited sources. The technique can also be used to create pseudo-polyphony within one voice or instrument to give the impression of multiple concurrent parts. Composers in the Baroque period such as Telemann,[111] with pieces like his *Sonata for Recorder in C major, (TWV 42-C2)* and Bach with pieces such as *Partita for solo flute in A minor, BWV 1013* frequently played with pseudo-polyphonic textures in solo instrumental works where rapid alternation between a high and a lower register gives the effect of two interleaved melodic lines with a convincing degree of separation.

Interlocking techniques were also elegantly employed by Schoenberg and Webern with the technique of *Klangfarbenmelodie*, Schoenberg in particular with the third of his *Five Pieces for Orchestra (Op. 16)*, and Webern with his widely known orchestration from 1935 of the six-part fugue *Ricercar a 6* from *The Musical Offering (BWV 1079)* of J. S. Bach. Unsurprisingly, given what has been said earlier in this volume about Steve Reich's compositional focus on repetition, the composer was directly influenced by the interlocking techniques of African music and used the term 'resultant patterns' to describe the emergent effects generated in his own process-based compositions.[112] Ligeti was also known for his admiration of African music and acknowledged that he was also directly influenced by it (incidentally, this influence is passed on to his son Lukas, who has worked extensively with musicians in Burkina Faso). In the 1980s Ligeti was already familiar with the research of Simha Arom on the music of Central Africa,[113] as well as the theories of Gerhard Kubik on inherent rhythms,[114] all of which provided substantial inspiration for his compositions such as his piano etudes *Fem* and *Der Zauberlehrling* (numbered 8 and 10, respectively) amongst many others in

---

[108] Stepputat and Samapati, "Performing Kecak."
[109] Jordania, *Why Do People Sing?*; Grauer, *Sounding the Depths*.
[110] Velitchkina, "The Role of Movement in Russian Panpipe Playing."
[111] Georg Phillip Telemann (1681–1767) was a German composer of the Baroque period, ranking among the most prolific documented composers in history.
[112] Reich and Hillier, *Writings on Music, 1965–2000*.
[113] Arom et al., *African Polyphony and Polyrhythm*.
[114] Ibid.; Kubik, "The Phenomenon of Inherent Rhythms in East and Central African Instrumental Music."

the 1980s and 1990s. Ligeti was fascinated by what musicologist Scherzinger called its 'psychological doubleness'[115] and in certain works he sought to achieve a kind of illusory musical space, as in *Continuum*, for harpsichord, having been inspired by such music as that of the royal courts of Buganda and the Banda Linda horn ensembles. Other specific works worth mentioning for sustained use of interlocking techniques include Louis Andriessen's[116] *Hoketus* for mixed ensemble 1975–1977, Charlemagne Palestine's[117] *Strumming Music* (solo piece for piano, 1974), Meredith Monk's[118] *Hocket* for two voices, Michael Gordon's[119] *Timbre*, and *Pléïades*, for six percussionists (1978) by Xenakis.

Still more striking examples of pseudo-polyphony are found culturally dispersed throughout the world, and outside of the official music institutions where the foregoing composers plied their trade. These include some cultural forms which have been rehabilitated from a kitsch status that some more open-minded critics dismiss as undeserved: for example, Bart Plantenga's multiple studies on the subject of yodeling, *Yodel-Ay-Ee-Oooo* and *Yodel in Hi-Fi*, provides a window onto the technical intricacies of a form that has expanded far beyond its origins in Switzerland, Bavaria, and Austria.[120]

Similarly, the practice of vocal percussion known as 'beatboxing' has broken through a kind of cultural novelty status to eventually be embraced by cultures far removed from its originary points in the boroughs of New York City. For example, contemporary YouTube channels such as SwissBeatbox (which has become the largest online beatbox platform since its inception in 2006), showcase the degree to which talented beatboxing artists are dispersed throughout the world. The level of creativity and ingenuity on display is summarized in the yearly championship Grand Beatbox Battle, which encourages the global beatbox community to push stylistic boundaries and continue innovating.[121] Finally, the *katajjaq* or vocal games of the Inuit are another example of a traditional practice that makes use of tightly interlocking vocal sounds to create deliberately ambiguous phantom results (we will discuss the *katajjaq* in much greater detail in the chapter *Phantasma Humana*).

2.1.4 Noise

As mentioned previously, we all naturally operate as apophenic creatures, constantly creating meaning from apparently meaningless, formless, or indistinct fields of

---

[115] Scherzinger, "György Ligeti and the Aka Pygmies Project."
[116] Louis Andriessen (1939–) is a Dutch composer of works such as *Workers Union, De Stijl, De Materie*.
[117] Charlemagne Palestine (1947–) is a visual artist and composer, specializing in lengthy 'overtone'-driven pieces for keyboard instruments.
[118] Meredith Monk (1942–) is a composer, filmmaker, dancer, and vocalist known for use of extended techniques in solo and group performance.
[119] Michael Gordon (1956–) is an American composer identified with the founding of the Bang On A Can new music ensemble.
[120] Plantenga, *Yodel-Ay-Ee-Oooo*; Plantenga, *Yodel in Hi-Fi*.
[121] www.swissbeatbox.com. Accessed October 26, 2019.

perception. In my compositional work, I have intentionally generated and used noise in such a manner, understanding it as a preeminent factor for the induction of sonic phantoms. In both my *Harp Phantoms* and *Drawing Phantoms* series of compositional works, there is a deliberate production of a perceptual environment characterized by substantial amounts of noise, whose aim is precisely to give rise to an enveloping generative field for sonic phantom induction. This strategy, which we could readily call 'noise field generation,' is in fact present—implicitly or explicitly, consubstantially or incidentally—in a wide variety of musical practices, from the overwhelming and all-embracing onslaughts of electronically amplified 'noise-as-music' of contemporary sub-cultures to the *mbira* thumb pianos of Sub-Saharan Africa (which are modified with bottle caps and other objects contributing to a 'noisy' timbral character).

Media theorist Friedrich Kittler[122] posited in his theoretical opus *Gramophone, Film, Typewriter* that "the conception of frequency undermines the privileged and distinctive status of written music and changes the concept of sound and music altogether."[123] This speaks to an evolving set of tendencies within the audio avant-gardes of the last hundred-odd years, particularly the divergent practices that are haphazardly labelled as 'noise' out of a kind of taxonomical frustration. It does have to be emphasized that, musically speaking, noise refers as much to a sort of creative *attitude* guiding an [anti]-genre as it does to sets of sensory data whose complexity defeat attempts at discerning a coherent message (and contrary to popular misconceptions about the exclusively transgressive character of noise, plenty of its adherents ascribe a therapeutic function to it as well).[124] Many examples of noise-as-art or noise-as-music knowingly play upon this fact that noise has both an objective/information theoretical meaning and a subjective/aesthetic meaning, with one prominent example being the Fluxus artist Yasunao Tone's *Musica Iconologos*. This is a piece in which digitally scanned texts from the *Shih Ching*—the earliest known collection of Chinese poetry—have their binary code converted into histograms and subsequently into sound waves, making for a sonic experience in which the formerly visualized text is no longer received "as message, but as sound, which is simply an excess."[125] Roughly the same de-semanticizing conversion process was applied later to *Musica Simulacra*, an overwhelming sound installation in which each poem in the Japanese *Manyo'shu* poetry anthology was rendered as a noise miniature.

Interestingly, there has become a growing attraction to noise-as-music because of, rather than in spite of, its lack of semantic transparency, while the purely functional quality of noise (i.e., utilizing white noise to test electronic circuits and to determine

---

[122] Friedrich Kittler (1943–2011) was a German literature and media theorist, influential for his role in the convergence of these two fields.

[123] Kittler, *Gramophone, Film, Typewriter*.

[124] Though not the only example of such, Helmut Schafer's thoughts on the subject here suffice to sketch a general outline of this attitude towards noise: "Noise can be seen as the chance to go into the back of things, as a therapy which is able to confront ourselves with others' lesser good looking sides of life without getting frustrated by this. See it as a source of very special energies, it gives power to see more clear and swim against." Schäfer, "Kraków, 05.03.05 | AudioTong."

[125] Tone, *Yasunao Tone*.

the acoustic signature of a performance space) has been expanded to include other kinds of psycho-physical 'tests' willfully entered into by listeners. A purely scientific/psychoacoustic definition of noise still goes a long way towards describing its appeal as a whole aesthetic:

> The interpretation of complex sounds which cover the whole frequency range is further complicated by psychological effects, in that a listener will attend to particular parts of the sound, such as the soloist, or conversation, and ignore or be less aware of other sounds and will tend to base their perception of loudness on what they have attended to.[126]

Put another way, this subjective focus upon one particular part of a sound structure, which may be entirely different from what the next person in the same listening space encounters, closely links the 'noise' listening experience to that of apophenia. Whereas more clearly tonal compositions would prompt the listener to experience them as an undifferentiated whole, a 'noise' composition of sufficient complexity would hold out a greater possibility for focusing on a constituent part that comes the closest to being personally relatable, be that the part played by a single instrument or a certain band of frequencies within the overall mix. An illustrative example here comes from Alvin Lucier's recollection of becoming inspired by his colleague James Tenney's[127] *Analog # 1 (Noise Study)*, whose filter-driven fluctuations of white noise were inspired by daily drives to New Jersey from Manhattan via the Holland Tunnel. Lucier, who had experienced the same sonic environment while driving en route to Pennsylvania, heard

> ... two layers of sound: one was a mid-range to high swishing sound, probably made by automobile tires; the other was a lower continuous sound, made by the motors. Both strands of sound were continuous and undulating. I got the image of an enormous flute being played by the fans that blow the bad air out of the Tunnel.[128]

A great number of so-called 'noise' compositions and improvisations rely upon engulfing the listener in a rolling sea of audio data with no clear meaning ascribed to it so that listeners may, like Lucier, complete the picture themselves. In fact, this has historically been seen as one of the positive distinguishing elements of composing or performing with 'noises' rather than with clearly pitched tones: a tutelary figure no less eminent than John Cage argued, in his *Lecture on Nothing*, for their creative use precisely because they "had not been intellectualized... the ear could hear them directly and didn't have to go through any abstraction about them."[129] Given what has

---

[126] Howard and Angus, *Acoustics and Psychacoustics*.
[127] James Tenney (1934–2006) was an American composer involved with numerous innovations in the realms of sound synthesis/computer music, microtonality, algorithmic composition, and sonic collage.
[128] Lucier, *Music 109*.
[129] Panzner, *The Process That is the World*, 114.

been said before about the brain as a prediction or inference engine, this is particularly interesting in the sense that such a lack of inherent meaning would make pareidolic activity almost unavoidable in listeners who otherwise subside on a diet of more clearly structured music. Key structural elements include wordless distorted vocals (or verbal content that is rendered incomprehensible by other elements in the noise environment), hectic collages of 'concrete' recordings, and lengthy passages of aleatory electronic feedback and segments that replace the musical emphasis on pitch and tonality with a focus upon timbral/textural differences as the primary communicative element.

Breaking this practice down into such constituent elements, while it may aid compositional novices in crafting their own noise-based compositions, might not do much to dilute its apparent hostility to uninitiated listeners. In his cybernetic opus *Steps to an Ecology of Mind*, Gregory Bateson offers a conception of noise that may therefore prove more useful in understanding the growing fascination with it than some alternative explanations; that is, the contrarian or antinomian impulse to be seen liking sound that is generally not understood, let alone liked. Bateson's thoughts upon noise as a creator of 'new forms' posits noise as an originary Ur-phenomenon of which more clearly defined arrangements of sensory data are epiphenomena (this theory is also echoed in the theorist Paul Hegarty's treatise on noise, in which he observes that "noise offers something... like dark matter, which may be what allows a structure for everything else to exist").[130] In suggesting this, they lend credence to the idea that participation in noise subculture is often an exercise in creating individualized meanings through pareidolia. In some cases, that sense of engagement comes about as an automatic reflex, much in the same way that Robert Ashley's[131] seminal noise-based composition *Wolfman* (1964) has been described ("the sounds are so powerful that you are in a continual state of analysis, your mind constantly moves in an effort to isolate the minutest details").[132]

Fascinatingly, this precise 'state of analysis' conjured by Ashley's piece cleaves close to the realm of more clinically oriented psychoacoustic experimentation, and to the identification of sonic illusions, which is often the intended result of this experimentation. A description of Cramer and Huggins's 'Huggins pitch illusion' seems as if it could be describing both the construction and sensory impact of a noise-based recording or performance experienced outside of enforced laboratory conditions:

> In the Huggins pitch demonstration, the sounds delivered to each ear are white noise signals with the exception of three narrow bands; 400 to 440 Hz, 500 to 550 Hz, and 600 to 660 Hz. Both signals begin in phase, and then the phase of the signal delivered to one of the ears is advanced by 180° in each narrow band successively. The perceptual impression is that the noise is accompanied by

---

[130] Hegarty, *Noise/Music*.
[131] Robert Ashley (1930–2014) was an American composer of operatic, theatrical, and electronic-based works, and director of the ONCE multimedia festivals in Ann Arbor during the mid-1960s.
[132] Lucier, *Music 109*.

a tone with gradually increasing faint pitch. No pitch is heard when the left or right ear signal is heard alone. Therefore, the Huggins pitch is a result of binaural processing of two slightly different signals.[133]

To be sure, there are more than enough creators within this community who steer listeners' interpretations via narrative or thematic cues, but a 'hands off' attitude towards priming audiences' expectations, and the subsequent encouragement of pareidolia, seem to occur here with a frequency much higher than that seen in other genres of music. This can be partially traced to the Cageian ethos of "letting sounds be themselves," a concept closely related to Cage's adoption of the Zen philosophy and shared with likeminded contemporaries such as La Monte Young (in a retort to those who felt this attitude might be "too extreme," Young once asked "do you think sounds should be able to hear people?").[134] This is also an echo of Hermann von Helmholtz's[135] affirmation that "tones and the sensations of tone exist for themselves alone, and produce their effects independently of anything behind them."[136] The application of this instruction to both composers and listeners has resulted in an increasingly fertile landscape of sonic possibilities, though not without proposing some severe challenges when imaginatively attributing sounds to something other than their obvious source: for example, Trevor Wishart suggests in his *Sonic Art* that "the sound of a bird need conjure up no ... metaphorical association [to flight]."[137]

All of this conspires to make noise perhaps the most natural and immediate candidate for the induction of sonic phantoms. This might be an intriguing or challenging statement for some, but this would be so probably only because of the mainstream conception of noise as something like 'occasional undesired sound.' In stark contrast with this limited view, the past few decades have seen noise undergo a drastic and profound transformation of status that has propelled its role and significance to—among other things—a 'generative matrix' of all sorts of potential contents and effects. This view interprets noise not as a contamination or corruption of the signal (in the classic paradigm of formal Information Theory) but rather as a comprehensive field that contains—undifferentiated and simultaneous—all the possibilities, all the frequencies, all the colors, all the values.

This change is not only aesthetic, but also ethical, political, and ontological. From our current perspective, we could see this as a revolutionary shift, but it is important to realize and acknowledge how much of the essence of such a perspective is naturally integral to other cultures (and has been so for a very long time). The epitome of this is probably the Kaluli people of Papua New Guinea,[138] who experience the broadband noise of waterfalls as a natural all-embracing source that potentially

---

[133] Scharine and Letowski, "Auditory Conflicts and Illusions."
[134] Young, "Lecture 1960."
[135] Hermann von Helmholtz (1821–1894) was a German theorist famed for his advances in theories of vision and acoustics, as well as for contributions to the philosophy of science and aesthetics.
[136] Helmholtz, *On the Sensations of Tone*.
[137] Wishart, *On Sonic Art*.
[138] Feld, *Sound and Sentiment*; Feld and Institute of Papua New Guinea Studies, "Music of the Kaluli."

contains all melodies of human music. As we know, this would run strikingly close to an understanding of white noise (and other versions, like 'pink' or 'brown' noise, or any other form of broadband noise) as a universal simultaneous container of all sonic frequencies, and therefore a generative matrix of sonic possibilities.

2.1.5 Accentuation

We could use the term 'accentuation' to encompass the numerous strategies and techniques to intentionally highlight, expose, or emphasize specific sound elements or features from a relatively undifferentiated sonic field or ground. This is somehow the inverse—or, perhaps more precisely, the complementary process—of the aforementioned 'noise field generation.' That is, once we have generated a rich and fruitful background field, we actively bring out and reveal sonic phantoms as a result of a purposeful and conscious procedure. *Actively* is, of course, key here, since this kind of strategy implies a more 'artificial' generation of sonic phantoms: not only do we set up a process of induction, but also compel perception to reach an apophenic state.

In technical terms, this is accomplished with relatively simple and fairly common techniques that would not be beyond the capabilities of any home recording studio, for example, equalization of either a live or pre-recorded audio signal. For those unfamiliar with this essential signal processing practice, equalization is the somewhat misleadingly named process of selectively spotlighting or foregrounding certain frequency components (i.e., *not* making all possible frequency components 'equally' perceptible): this is done by boosting or reducing the selected components to achieve a final mix to the tastes of the producer. In this sense, it is not to be confused with filtering, which is a method of simply subtracting frequency components entirely, though the two processes are by no means mutually exclusive.

This can also be achieved with other specific reinforced actions on instruments or objects to attain the desired level of emphasis for the sonic phantoms to become patent. In many disparate music realms, from so-called 'acousmatic music' to 'electronic dance music,' and numerous other forms in which the mixing desk itself becomes the primary 'instrument' for live performances, equalization controls are indispensable as means of providing dramatic emphasis, as well as creating dynamism from passages of sound that might, for one reason or another, otherwise be perceived as static. Using equalization to timbrally 'brighten' a bass-heavy, chthonic passage can produce perceptive changes in spatiality or temporality that are disproportionately large compared with the small amount of effort required to do so. This use of equalization as a timbre-manipulating device takes the practice more clearly into the aesthetic realm and out of the purely functional one (equalization was originally conceived of to help with correcting frequency response errors that occurred during the recording process, e.g., compensation for microphones' directional relationship with frequency). This is, of course, a form of purposeful manipulation of the classic 'figure/ground' perceptual situation. Many possible levels of intuitive attention shifting between sonic planes or phases are deliberately tweaked to modify what, in normal conditions, would be the immediate and expected figure/ground relationships of perception.

Such conscious and intentional reinforcement of specific sonic features in the music with the aim of generating sonic illusions also take place in music far removed from electronically enhanced processing. This can be witnessed, for example, in the amadinda or akadinda xylophones of Uganda: phantom patterns that arise in the music can be forcefully accentuated, after becoming hearable, by the musicians who caused them or by specialized additional musicians who specifically reinforce these phantom patterns with their playing.

As described in the following chapters, I have extensively used various forms of accentuation in my work to generate sonic phantoms, including the aforementioned means of equalization and filtering of sonic materials (in both live and in studio recordings). I have also actively highlighted specific sound features—and thus generated sonic phantoms—by analyzing certain instrumental actions and consequently providing precise instructions for performers to dramatically emphasize their actions and hence the sonic results.

2.2 Realms of Sonic Phantasmatic Experience

Having exposed the fundamental techniques that contribute to the realization of sonic phantoms, we now turn to introduce a contextual framework that we propose to encompass whole realms in which these sonic phantoms can be manifested and experienced from a compositional perspective.

Anicius M. S. Boethius, a philosopher of the late Roman-Byzantine Empire, was a polymath instrumental in developing correspondences between what he posited as the *quadrivium* or four great arts of arithmetic, music, geometry, and astronomy. Each of these subjects was intended to be assessed in the precise order listed, with knowledge of the prior subject being necessary for understanding of the next and, ultimately, for a reasonable grounding in philosophy (which itself was laid out in his *De consolatione philosophiae* [The Consolation of Philosophy]). Boethius would be highly influential upon later thinkers by describing the Pythagorean unity of mathematics and music, laying the early groundwork for ambitious mathematically inspired compositional ventures from Bach's *Well-Tempered Clavier* to Iannis Xenakis's stochastic pieces. Thanks to Boethius's system, music was also primed to become an essential component of Renaissance education. Meanwhile, much of his life was dedicated to the translation of the original Aristotelian and Platonic works into Latin, with other representatives of Greek antiquity contributing to his landmark works such as *De Arithmetica*, which was subsequently popularized during the medieval period of philosophy.

Boethius also summarized ancient Greek thought on music in his *De Institutione Musica* [The Principles of Music], written at the end of the fifth century. It, too, would have a profound influence on the philosophy of future intellectual eras, eventually becoming one of the first musical works to be printed in Venice in the fifteenth century. Renaissance polymaths such as Robert Fludd would later take much interest in its fundamental assertions, with Fludd in particular using Boethius's comprehensive tripartite system for music classification as the jumping-off point for his own far-ranging discussion of 'macrocosm and microcosm'. Marin Mersenne, in his own noted music theory work *Harmonie Universelle*, was inspired by Boethius's sense of integrality or of a possible 'universal language' that could spring from the proper unification of rhythms and harmonies, and eventually improve the moral standing of those equipped with this knowledge. As such, *De Institutione Musica* constituted a conceptual framework for the very concept of music and its practice that remained in place in the Western world as the main reference for musical understanding for over a millennium. This system considered the following categories of music:

— *Musica Instrumentalis*: instrumental music.
— *Musica Humana*: the internal music of the human body; human body and spiritual harmony.
— *Musica Mundana* or *Universalis*: the Pythagorean 'Music of the Spheres'; a metaphysical principle embracing the non-human philosophical world/cosmos as generator of an imaginary harmonic music caused by the movements of celestial bodies.

This was a radically inclusive and universalizing system; one that determined that the organization of music making expanded beyond mere classification to recognize that there is also a level of 'music' beyond what can be produced by the human body or by the instrumental extensions thereof.

When considering the breadth of my multiple compositional series *Sonic Phantoms*, I have followed the suggestion proposed by Francisco López to use Boethius's categorizing/framework system in an open way, to the extent that it is inspired by such universalizing openness in the acceptance of music—and related aesthetic/perceptive phenomena—as being potentially generated in realms other than the more traditionally musical. In relation to my own creative practice, however, this suggestion does not so much stem from the interest in the classification system *per se*, or in the Pythagorean-Boethian tradition *as is*, but rather in the comprehensive vision of which these systematizations are representations: an ethos that expands the 'musically possible' beyond the instrumental and the anthropogenic. Thus, for example, the *Musica Mundana*'s 'Music of the Spheres' is a compelling and inspirational historical example, in the context of Western philosophy, of a level that is beyond human intervention; music beyond the *compositional* reach of humans. In that spirit, the particular use here of the corresponding realm of *Mundana* is an expanded version that aims at embracing all of 'nature' (and not only the extra-terrestrial/cosmic), as we would commonly understand it today.

It was, in fact, as an a posteriori realization of my working process that a significant analogy was recognized between my personal concept of musical composition and the broad classification system of Boethius. I have composed a wide spectrum of sonic pieces with materials from diverse sources and origins: from drawing on amplified objects, to using text-to-speech software applications, to the choruses of frogs and insects in rainforests, as well as instruments and voice. To me, they are all equally musical. Fundamental to my work is an aesthetic and emotional conception of sound in which *Mundana*, *Humana*, and *Instrumentalis* would be equally valid as music; much as they were—in a different conceptual framework—for Boethius himself.

The four main sections of this volume thus structure the presentation of these *Sonic Phantoms*' compositional explorations according to four realms of sonic experience hereby called *Phantasma Instrumentalis*, *Phantasma Materialis*, *Phantasma Humana*, and *Phantasma Naturalis*. Each section explores the territory demarcated by these terms and focuses on a compositional series that manifests the phenomenon of sonic phantoms in one of these realms. Additionally, each section or sonic realm highlights particular uses, methods, processes, or implementations of phantom induction in my own music, contextualized by relevant ideas and inspirational research from a wide variety of sources, which are discussed in relation to each particular compositional series.

Phantasma Instrumentalis would probably be the most immediate category in a traditional musical sense, given our natural inclination to identify music as a human construction that has advanced technically and aesthetically through the use of tools or instruments, and thus is the only one directly equivalent to a Boethian category. This realm is represented by the project *Harp Phantoms*, a collection of compositions for 'prepared' amplified harp encompassing both live performances and studio-manipulated pieces. Notwithstanding its clearly instrumental nature, this project reveals, among other things, relatively unexpected outcomes for an instrument (at least in the traditional sense), such as the generation of noise fields and voice-like sonic phantoms.

Phantasma Materialis is a realm introduced to account for the significant and extensive body of work developed, for many decades now, in contemporary sonic practices dealing with the inherent musicality of objects. It basically refers to objects without a preconceived musical use that have ultimately become 'instrumental' in an extended musical sense. This realm is exemplified by my project *Drawing Phantoms*, an amplified, trance-inducing drawing ritual involving paper, writing utensils such as pencils or crayons, and a supporting, resonating surface. This work directly and contextually relates to research and artistic practices harvested from the creative—and, more generally, cultural—fields of trance and ritual.

Phantasma Humana would essentially refer to the more straightforward and specific—rather than the Boethian mystical or esoteric—expression of human bodily produced sound represented by the voice. This realm is exemplified by my project *Vocal Phantoms*, which centers on the generation of sonic phantoms using the human voice (both organically produced and synthetically approximated) as well as human language, as starting points for core compositional material. As part of the development of this project, I carried out research on both the vocal techniques of the Inuit *katajjaq* and Georgian polyphony, in expeditions to Baffin Island in Nunavut (2010) and to Georgia in the Caucasus (2012).

Phantasma Naturalis is the proposed realm for all sonic phantom phenomena generated from 'nature' in a contemporary philosophical sense; that is, the non-human or non-artificial territories, whether cosmic or terrestrial. Again, in this case the focus is placed more squarely upon perceivable sonic entities than upon the 'mystical spheres' that would have been a characteristic object of study in Boethius's time. *Phantasma Naturalis* is represented by a series of recordings and compositions, under the title of *Natural Phantoms*, which explore the phantom-producing polyphony of different natural environments. These sound pieces have been created from original source materials recorded in several field recording and listening expeditions that I carried out, in partnership with Francisco López, to rainforests and other natural environments in Brazil, Borneo, Cambodia, Australia, Bolivia, Chile, and South Africa between 2011 and 2019.

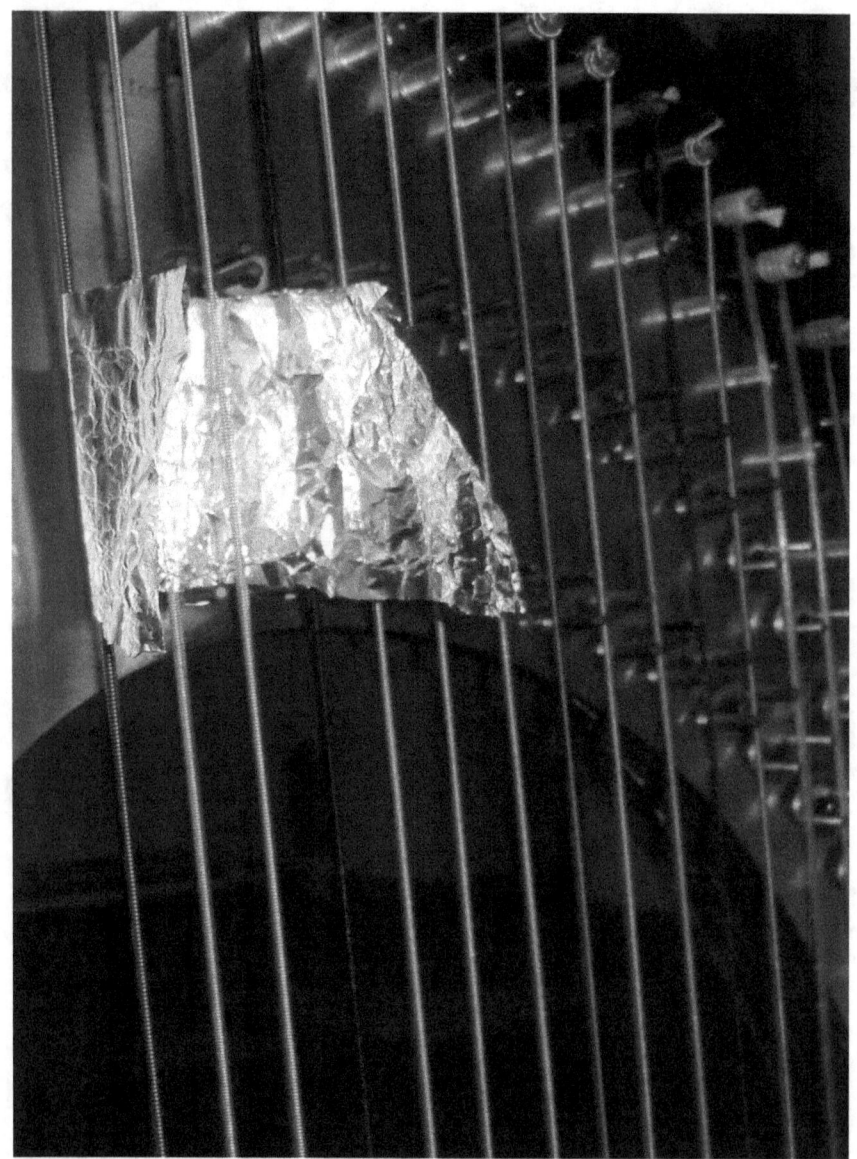

Figure 3.1 *Harp Phantoms* compositional series. Foil in between harp strings. Photograph by Barbara Ellison, 2011.

3

Phantasma Instrumentalis: The Realm of the Instrument

Compositional Series: *Harp Phantoms*

3.1 Sonic Exploration of the Instrument

As mentioned above, there are many different approaches and techniques for obtaining sonic phantoms in the realm of the instrumental. In this chapter I will discuss my own use of interlocking techniques built from modest and simple starting points, coupled with other techniques such as layering, in order to form intricate, dense, evocative, and hypnotic structures with phantom effects of pseudo-polyphony.

I have experimented over the years in many different pieces with various instruments exploring sonic phantoms in the realm of the instrumental. The compositional series of pieces I refer to collectively as *Harp Phantoms* constitutes a good example of my in-depth efforts in this direction, having evolved from an extensive period of dedicated exploration with a 'prepared' concert harp, that is, one making use of both point-local and overall-ambient amplification.[1] These pieces evolved from a methodology of action-based process work, and were thus developed working in a very direct and unorthodox way, using the harp as a complex sound generator and resonator.

A large part of the practical process carried out in this compositional series involved making an in-depth exploration of the harp itself, including the ability to radically alter its tonal and timbral qualities via extended technique or 'preparation,'[2] in various configurations and setups. From the onset of this process, I was keen to discover in a practical way the under-explored sound-making potential of this instrument, extending its sound potential beyond its historical and more traditional musical context. Though the instrument has not been absent from music that adopts a challenging or avant-garde posture (e.g., Zeena Parkins's contributions to the New York

---

Readers can listen to several pieces of this compositional series on the website www.sonicphantoms.com

[1] See following note on 'preparation.'

[2] The concept of the 'prepared' musical instrumentation, inaugurated in the twentieth century and popularized after early efforts by Henry Cowell and John Cage, refers to instrumentation that is modified in such a way as to drastically alter tonal qualities beyond what the instrument is otherwise capable of. For example, the placement of foreign objects or 'preparations' on the strings of a piano essentially converted it from a generator of harmonies and melodies to an instrument capable of sharp percussive attacks.

City 'downtown scene' of John Zorn, Ikue Mori et al.), this repertoire is not extensive, and the novelty of deploying it in unorthodox contexts makes it all the more potent as a generator of sonic phantoms. In my own work with the instrument, I began with a hands-on creation of all kinds of sonic blotscapes, to trigger the sonic imagination, a path that seemed more rewarding than thinking up the sounds in advance, notating or preparing them, and then getting the player/performer to produce them.

My explorations then led me to develop a stockpile of preparation techniques and gestures, which gave rise to varying sets of textures and processes. The compositional material was ultimately derived from the sonic blotscapes generated by the prepared instrument/object/player/bistable system. Whilst developing the pieces at this stage, I was fortunate to be able to work directly with harpists Angélica V. Salvi[3] and Rhodri Davies,[4] both of whom devoted much time and attention to the exploratory process. We spent many hours together workshopping the possible playing strategies and experimenting with adaptations to these strategies made through practice. These techniques will always depend to some degree on a level of adaptation, responding to factors such as the acoustics of the space, instrument tuning and preparation, performing musicians, and time available for the performance itself.

The compositional series *Harp Phantoms* evolved from an initial semi-scored piece for live performance, which in turn emerged from a personal gestational period of practical, hands-on research with the harp. Formally, the first piece is comprised of ten 'scenes,' with each scene focused on exploring a specific action or process. Each of these scenes also constitutes a piece in itself that can be performed as such or in combination with another scene from the series. When employing devices very like those just mentioned, each of the respective scenes takes on a kind of ritualistic character, that is, one in which an enhanced level of focus or concentration is adopted to increase the possibility of unanticipated and personally revelatory results.

More specifically, each scene within the total series focuses mainly on one specific action or behavior and is defined in as economical a way as possible, that is, on a single page of the score. Each action is expressed textually by a series of directions or instructions along with accompanying diagrams and by sound and video examples. Through the process of experimenting with playing styles that favored raw physical gestures on the harp, and by playing with intense repetition, I discovered different ways to create a variety of sounds and textures, ranging from the very noisy (by means of scraping) to the harmonic (using bits of Blu Tack adhesive material, the blue artificial clay-like material used, for example, for attaching posters to walls).

The actions themselves produce cycles of repeated patterns, ranging from simple interlocked two-note *ostinatos* to repetitive *glissandi* gestures. Many of these pattern-producing actions make use of 'dovetail' interlocking techniques (explained below). The actions were devised, constructed, deconstructed, and ritualized through the initial practical experimentation stage, resulting in a working set of instructions, which aimed to reduce the actions to their bare essentials. In a sense, the actions themselves, because of their repetitive nature, function as a network of looping structures.

---

[3] Angélica V. Salvi, https://vimeo.com/angelicavsalvi. Accessed October 26, 2019.
[4] Rhodri Davies, http://www.rhodridavies.com/. Accessed October 26, 2019.

3.2 Instrument Preparation and Sonic Blotscapes

The following is a brief description of the preparation of the harp for the *Harp Phantoms* compositional series, detailing the steps needed in order to transform the sound of the harp itself. A number of different techniques were devised and tested through a process of trial and error during intensive practical workshops.[5]

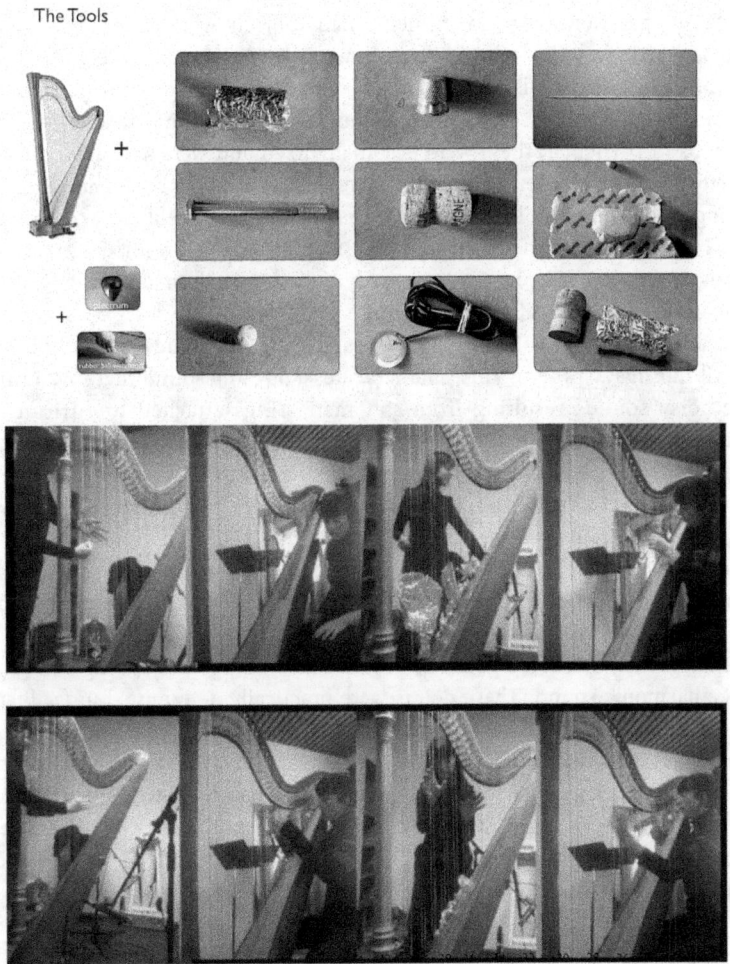

Figure 3.2 One of my own workshop sessions experimenting with different materials to 'prepare' the harp: foil, thimble, wooden stick, screw, cork, Blu Tack, contact mic. Image part of documentation of one of the scores of the compositional series *Harp Phantoms*. Image by Barbara Ellison, 2011.

---

[5] I was fortunate to have individual and, most importantly, unsupervised access to a concert harp (and a sufficiently generous amount of time to make use of it) at the Royal Conservatory of The Hague.

Whilst instrumental preparations have been extensively explored in relation to the tradition of the 'prepared piano,' their use with the harp in general is only beginning to be more widespread. That said, harpist Rhodri Davis's exceptional practice involves extensive use of various objects and materials for harp preparations. The aesthetic of buzzing sounds was integral to the timbre of these pieces (also inspired by similarly noisy timbres of African harps and *mbiras*). Indeed, most of the techniques and preparation strategies that I followed, described below, were selected because of their capacity to generate what I have previously described as sonic blotscapes.

In the preparation of the instrument, many objects were added to the harp as effectors, which would work to intensify sonority or increase the noise-to-pitch ratio. The objects eventually selected, after some trial and error, were aluminum foil rings and sheets, Blu Tack blobs, champagne corks, screws, and sticks. As to the Blu Tack blobs in particular, these were eventually fixed and were activated, directly or indirectly, by plucking the strings. This material functions to create harmonics when the string is plucked, as well as detuning the harp strings. As a deliberate compositional choice, I wanted to alter the tuning of the instrument without resorting to traditional methods of doing so (i.e., having to actually tighten or loosen any strings), and this technique worked very well for that purpose.

The foil rings were attached to strings at different points where the strings could vibrate against them. This, again, creates a buzzing sound that is reminiscent of the noisy sounds resulting from soft metal rings attached to African *mbira* instruments, which in this tradition are believed to aid in the communication with the spirits of the ancestors.[6] These kinds of preparations function as noise field generators, as well as timbral transformers, creating extra sonic layers from which sonic phantoms arise.

The score of *Harp Phantoms* itself includes necessary graphic and descriptive information that is, in fact, a documentation of the process of exploration and 'preparation' of the instrument. Blu Tack blobs were attached and tacked to the harp strings at specific, measured points to produce a detuning of the string and to give a bell-like harmonic sound. These points were practically determined to facilitate the bell-like sounds when plucked. Measurements of the Blu Tack placement are indicated in the score from the bottom of the string at the soundboard to a position in the middle of the string. In reality, whilst these measurements were precise for the specific instrument I was working with at the time, I discovered that different harps require some adjustments: a bit of playing and testing the precise positions to attain the best sound to suit each particular string. Aluminium rings were placed loosely around certain harp strings to create noisy buzzing artifacts and rattles when the string is plucked and scraped with a plectrum. Foil pieces were also loosely placed and threaded around and between the strings for noisy timbral metallic effects.

Champagne corks were also wedged in between strings, for percussive effect. At different moments in time during the performance of the piece, they are beaten softly with a stick attached to a rubber ball, creating a low resonant, drum-like sound.

---

[6] See Berliner, *The Soul of Mbira*.

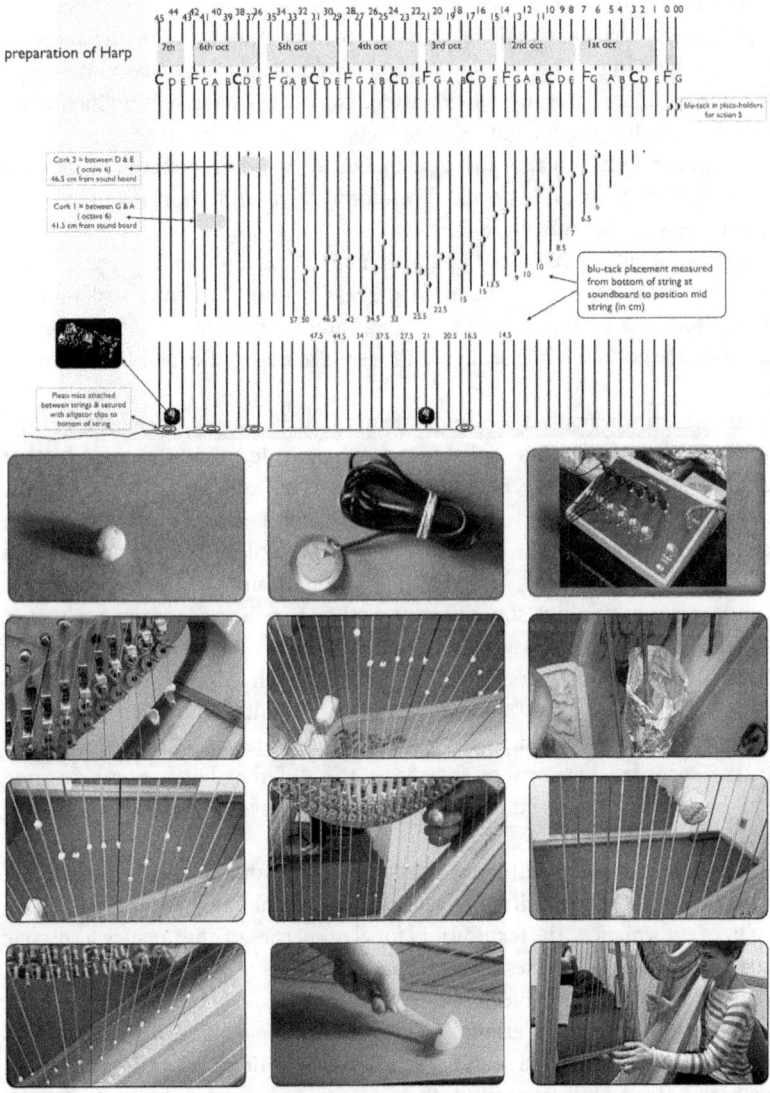

Figure 3.3 Diagram and photographic illustrations with harp 'preparation' for harpist Angélica V. Salvi's set-up tests: Blu Tack, contact mic, custom-made contact mics' mixer, objects placed on harp's strings. Image by Barbara Ellison, 2011.

Piezo-electric microphones ('contact mics') were used for local amplification of the prepared strings with objects. A minimum of five microphones were attached to strings near the base of the sounding board, where they unequivocally brought out the more subtle sonic details of all the resonances of the instrument. Such details would normally have gone unnoticed without this 'close-mic' style of amplification.

A thin wooden food stick was used as a performance tool to generate an 'in-between' action, consisting of a quick, repetitive, back-and-forth motion with the stick between the strings. When set in motion this way at the base of the string, a nice resonant sound can be achieved, which results from the rhythmic contact between the amplified wood of the stick and the harp string gut. A large metal screw was also used to produce a 'third-bridge guitar'[7] effect to alter string pitch. One hand holds the screw against the string as a guitar bridge, moving up and down for pitch control, while the other hand plucks the string with the wooden stick.

A mallet with a rubber ball head and a thimble placed on one finger were used in conjunction, although in alternation, as percussion tools to gently strike a cork placed in between the strings, whose sound was amplified by means of the contact microphones at the bottom of the strings. This results in a beautiful, bell-like, chthonic rumble.

A guitar plectrum was used to scrape the amplified C string of the 7th octave, producing a nicely noisy and granular sound, whose unique features were further enriched by buzzing foil attached around the string and amplified with the contact microphones. This started as a strong, isolated performance gesture with the plectrum, and then evolved in later versions of the piece (including recorded versions), by means of layering, to generate a wall of granular sound. Layering of these sounds was achieved by using different methods, such as a looping station (an analog effector with a built-in sampler) for live layering, and recording/playback computer software in studio versions of the piece. Though this plan was eventually tabled, I had also contemplated the use of multiple simultaneous harpists for this particular purpose. Timbral complexity was further enhanced by detuning different layers in the recorded versions of the piece: the different layers were detuned slightly higher or lower to each other to create a more complex and remarkable sonic texture.

I also experimented thoroughly with amplifying other common objects to explore their sound-making potential, with the eventual aim of combining them with the harp sounds in live performance. The textural and tonal properties of objects such as bicycle wheels, glass, and cardboard boxes with attached elastic bands were explored in depth. Using some of these prepared objects, I then created hybrid live versions of the piece for two or more performers; for example, a version for harp, amplified bicycle wheels and cardboard boxes. These extra sounds generated an additional noise field, as well as providing harmonic and buzzing sounds that uncannily matched and accentuated the output coming from the harp itself.

---

[7] Artists famous for using 'third-bridge' setups include Sonic Youth, Glenn Branca, and Fred Frith, following pioneers such as Harry Partch and John Cage. Yuri Landman (1973–) is an experimental instrument builder based in The Netherlands who has created special instruments based on prepared guitar techniques and has custom built derivatives for performing artists. I have myself also used his amazing instruments (www.hypercustom.nl).

3.3 Ritualization in Performance

As hinted at by the earlier discussions on musical techniques, one of the main features of my compositional work is the use of repetition, along with exaggeration, elaboration, and other forms of accentuation, to intensify and highlight patterns and phenomena that might seem apparently ordinary or indistinct. I find that the 'normal' can undergo a fundamental transformation that renders it more sublime and expressive than what it superficially appears to be: this assumption lies at the core of my practice; and I believe that it has to do, at a very essential level, with aspects of ritualization of performance.[8]

Ellen Dissanayake's theories of ritual and ritualization[9] in the context of art and culture have been inspirational, in terms of reflecting on my own use of ritualization of action and process.[10] She uses the term 'artification' to describe how an everyday normal action is 'artified' or 'made special' through a ritualization process, and consequently transformed, to "make ordinary reality extraordinary."[11] Some of the techniques that Dissanayake considers as 'behaviors of artification' are processes of repetition, regularity, stylization, exaggeration, and formalization. In my own process of sonic ritualization, I use equivalent structural features of a fundamental nature such as the aforementioned repetition, exaggeration, and elaboration of sounds, movements, and so on, to intensify and to engage the attention of others.

Even though some of the performative actions on the instrument are more easily repeatable than others, in general I tried to fine-tune them to make it possible to automate and maintain a regularity of repetition carried out over extended durations. When executing actions that can be repeated and internalized over time, the performer and the listener are both provided with a means of magnification, or a higher sensitivity to individual perceptible events: this in turn allows each party to zoom in and 'get into' or internalize the gesture, so as to magnify and reveal all its micro-details over time. Doing this throws a 'perceptual spotlight,' so to speak, on detailed aspects of the performative action, to reveal nuances that we normally would not notice. It is a process of intensification and elaboration; when the mundane action is made special by ritualization, it achieves the kind of significant transformation described by Dissanayake above: abstracted, heightened, elevated, and inducing a sense of awareness on its new significance.

---

[8] Practical research into ritual and performance has over time involved regular workshopping and performances with the 'Trickster' collective of artists (www.tricksterspace.org) of which I am a core member, and also undertaking other workshops, including participation in intense physical theatre workshops in London with Nicolás Núñez, director of the Taller de Investigación Teatral in Mexico City, acclaimed author and former collaborator with Jerzy Grotowski. See Nunez, Middleton, and Fitzsimons, *Anthropocosmic Theatre*.

[9] Dissanayake theorizes from an ethological perspective that animal ritualization and rituals in human culture (e.g., in music making) have a common evolutionary origin. That music, for example, is understood to be a 'behavior' that evolved in ancestral humans because it contributed to their survival and reproductive success. See Dissanayake, "Art as a Human Behavior"; Dissanayake, "Homo Aestheticus."

[10] Ibid.

[11] Ibid.

In addition to this process of ritualization, I specified for this series a range of extended performance durations (from half an hour to several hours). Through this combination of ritualization and persistence, I wanted to bring both performer and listener into an intensified trance-like state of complete immersion in the process. One of my primary aims as a composer and performer is, indeed, the creation of an immersive situation in which a blurring of distinctions between action and result (i.e., between performative movement and sonic output) is encouraged and facilitated, being itself the natural result of simultaneously performing and listening. Achieving a richly complex variety of perceptual changes and deviations over time, with only relatively simple gestures, is the intended reward of this process.

Once sonic patterns have been established and stabilized in the way that I describe, the performer is then tasked with the challenge of perfectly looping the pattern; repeating his/her actions as soon as a discernible pattern arises. This is done in order to provide a stable perceptual foundation or 'ground' (in the sense of the 'figure and ground' perceptual situation). In terms of the accuracy of the repetitions, the desired result might provide another sort of blurring effect, this time by projecting a mechanical-repetitive propulsiveness: imagine, for example, a pattern that has been recorded into a digital buffer and is stuck in replay or loop mode.

Upon the establishment of a stable looping ground pattern, the perceptual malleability of the sound should become more noticeable, with the brain's interpretive abilities engaging with the materials to playfully transform and mutate them prior to any actual physical elaborations of, or alterations to, this acoustic material. In reality, of course, it is extremely difficult to make flawless repetitions (i.e., ones that would be perceived as being completely identical from one to the next) without the assistance of technological or computational means. However, failure in this regard is not necessarily a failure to achieve the intended results outlined above: it is precisely the impossibility of complete accuracy, and the variance in rates of accuracy achieved, that often makes for very interesting results. The action type will, naturally, have great influence over this aspect, as does the ability of the player to seamlessly iterate. A key component of this process is, again, repetition: when this is applied to interlocking sonic patterns, the total sonic output can seem disproportionately complex relative to the simple input. As all of this unfolds in real time, attentive listening can tune into the subtle, initially imperceptible transformations of emergent patterns (i.e., patterns going out of phase), which can suddenly surface in perception and just as suddenly become submerged again.

For the player in performance, if new patterns emerge in perception and are then heard as phantom artifacts, they can reinforce these patterns and bring them clearly to the foreground through the accentuation process (emphasis and elaboration). This is, of course, a feedback process, a system in which the interrelationships and interactions between instrument/object/action/space at different levels of perception give rise to the emergent result. The resultant sonic image emerges from this network of relationships, from the constant perceptual alternation of this figure-and-ground bistable system.

3.4 Structure and Possible Variations

The score of *Harp Phantoms* functions as both a skeletal documentation of the process of development of the series and as a functional structural map of performance instructions (as is the case with many of my scores). It is divided into several scenes, which become each piece and can be performed in a set given order, with specified approximate durations. In different performance situations, these durations can be easily adapted, depending on the overall time set in advance for performance.

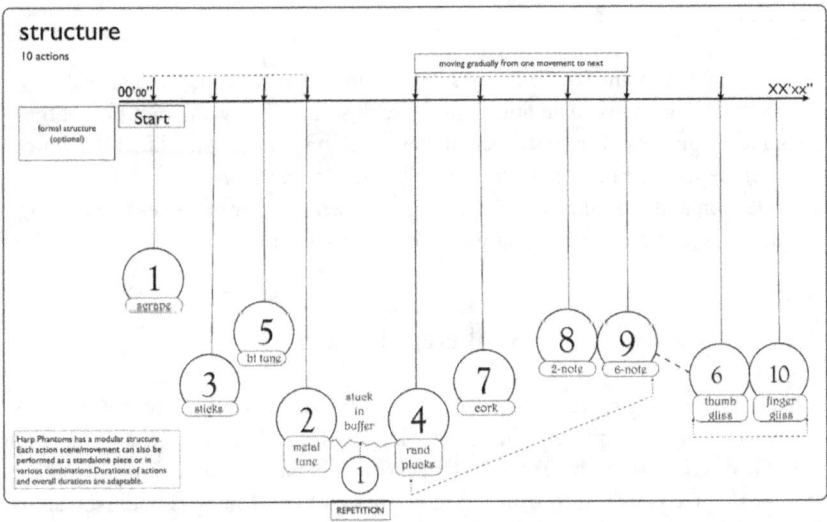

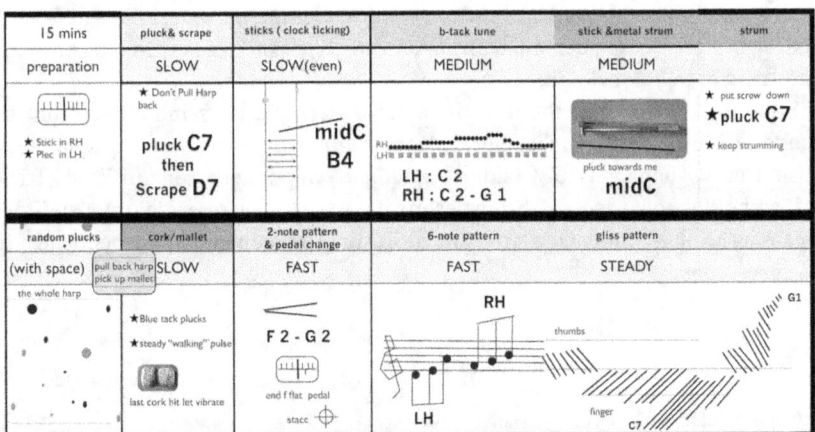

Figure 3.4 Part of the score for the compositional series *Harp Phantoms*, showing one version of a structural timeline with selected scenes and order of play for a live performance. Score by Barbara Ellison, 2011.

I intensively experimented, together with the harpists, with many variations of internal formal structure and duration of this series in performance. As this project has developed over the course of several years, the most interesting and successful *Harp Phantoms* performances for me were reductions of the series to only one or two scenes, combined together and extended in duration, sometimes over several hours. Such prolonged timeframes are conducive and necessary for entering into liminal or heightened states of perception by inducing a state of intensified absorption that will reveal sonic phantom patterns.

There are a number of approaches to performing this piece and all can be deconstructed, adapted, and played with; they can be summarized in two main categories:

1. 'Straight' performance of all action scenes in the series, in a set given order, with specified approximate durations for each scene. The overall duration of the combined individual scenes is dependent upon the performance situation, which in turn depends upon the overall time allocated for performance.
2. A selection and modular combination of a chosen number of scenes and playing order for specified overall duration of performance time.

3.5 Inherent Patterns

Both the stream segregation theory[12] and the figure-and-ground reversal theory[13] mentioned above conceptually illuminate what is actually taking place at a perceptual level when composing with these types of ambiguous interlocking structures. Additionally, and from a more distinctly ethnomusicological viewpoint, the research of Austrian musicologist Gerhard Kubik into what he has described as 'inherent patterns' and the auditory streaming phenomenon is worth considering here, as it significantly sheds light upon these phenomena from an extensive experience with a particularly rich strand of traditional music.[14]

Kubik first introduced the term inherent patterns in his seminal work on the composed illusory patterns that emerge from the *amadinda* and *akadinda* xylophone music from Uganda.[15] Whilst taking *amadinda* xylophone lessons in East Africa, and without knowledge of the experimental laboratory research into auditory streaming, Kubik stumbled across the streaming phenomenon independently in

---

[12] Bregman, *Auditory Scene Analyis*; Handel, *Listening*.
[13] Beardslee, Wertheimer, and Rubin, *Readings in Perception*; Gombrich, *Art and Illusion*; Arnheim, *Art and Visual Perception*.
[14] Kubik, "The Phenomenon of Inherent Rhythms in East and Central African Instrumental Music."
[15] Ibid.

the field.[16] Inherent patterns, as put forth by Kubik, can be defined as independent melodic–rhythmic phrases that only exist as an aural image and are not played as such by the performers. He described them as a Gestalt-psychological phenomenon caused by a certain structural arrangement of rapid passages with wide intervals.[17]

Personally, it was a revelation the first time I heard a recording of *akadinda* and *amadinda* xylophone music from Uganda, as it is such an extreme realization of an effect that I had myself been playing with and exploring for years with different instruments.[18] It is a truly impressive example of sonic phantom-producing polyphony induced by auditory streaming, and I share with Kubik the fascination and appreciation of the intricacy of these musicians in successfully manipulating the human perceptual system with such expertise. Kubik described the unique method through which the court composers of Buganda had "learned to play tricks on auditory perception,"[19] which involved composition of pieces exploiting the way in which rapid interlocking cycles of pitches with certain structural arrangements and wide intervals can facilitate the production of sonic phantoms. Moreover, Kubik noted that obtaining the resulting pattern *Gestalts* (which, for our purposes, are essentially sonic phantoms) is indeed the objective of the process; they are not just mere incidental artifacts of the faulty human auditory perceptual apparatus. The focus upon inherent patterns is "what many African composers are after by passion":

> There is a psycho-acoustical fact which African composers particularly of instrumental music (xylophone, likembe, etc.) are delighted to take advantage of: that the human mind is inclined to join together form objects of similar or equal qualities and establish a 'gestalt'. In music the listener associates notes of equal color or loudness and of equal or similar magnitude. If, further, notes of similar qualities are arranged in a definite rhythm of occurrence then association is enormously stimulated.[20]

---

[16] Unbeknownst at the time to Kubik and far away from field research in Africa, as early as 1950, researchers Miller and Heise were conducting experiments into the streaming phenomenon (Miller and Heise, "The Trill Threshold"). This type of seminal work promoted the later development of a substantial research focus in the 1970s in investigating parsing mechanisms in auditory perception. The experiments of Bregman and his collaborators have demonstrated a number of different acoustic dimensions that can be used by perception to group together the components of a complex sound. Extensive tests were carried out under laboratory-controlled conditions with cyclic repetitions of simple-stimulus sequences to analyze the grouping tendencies under certain conditions, as guided by the Gestalt laws of organization.

[17] Kubik, *Theory of African Music, Volume I*; Kubik, *Theory of African Music, Volume II*.

[18] The *akadinda* and *amadinda* are both large wooden xylophones played by several musicians at the same time. The resulting music consists of rapidly interlocking patterns of isochronous notes and disjunct intervals firing at about 8.4 tones per second, repeating the same note sequence in cycles for extended periods of time.

[19] Kubik, *Theory of African Music, Volume I*, 2.

[20] Ibid., 71.

When referring to the *amadinda* xylophone music, Kubik also highlighted the perceptual mismatch between the sensory input of the players and the resulting illusory 'sound image':

> By applying the economical technique of interlocking performance parts, and through an irregular structuring of pitch sequences in disjunct intervals, composers learned to create auditory jigsaw puzzles that would oscillate in perception.[21]

Listeners thus perceive patterns that are not played as such by any of the players, but are instead results of the perceptual restructuring mechanisms of the brain. In this fashion, the sounding image of the music often differs quite dramatically from the notes that we can see being played. I had experienced myself first-hand such an outstanding mismatching in the Inuit *katajjaq*, the vocal 'games' typically performed by two women in the Arctic (see chapter 5). All of these cases with clearly visible performers are blatant examples of 'what you hear is *not* what you see'; instead, what you hear is in fact the result of a perceptual restructuring of the normal brain operation which, as described by the theories of prediction error minimization,[22] is trying to make sense of the input. This is a fundamental core aspect of my work with sonic phantoms, and I somehow equate the compositional process of generating sonic phantoms with artistically exploring the active nature of perception itself.

Interlocking patterns permeate my own compositions in various ways, and help in contributing to the multilayered polyphonic structure of the music. The essence of this kind of pattern construction, linking it to the classic figure-and-ground bistable images, is that the patterns themselves share common borders with common elements. This sharing of borders causes confusion, as we attempt to organize the incoming auditory signals into coherent patterns, making perception ambiguous. In my compositional series *Harp Phantoms* the performer is able to generate a kind of indirect polyphony through the creation of instrumental patterns that, confirming the principles outlined by Bregman's Auditory Scene Analysis, split into subjective streams, thereafter producing a contrast between a 'played' and a 'sounding' image.

3.6 Auditory Streaming

The literature on our intrinsic motivations for being innately *compositional* creatures is vast and deep, though not difficult to comprehend. As neuroscientists Ramachandran and Hirstein have hypothesized, this innate drive towards synthesis of disparate

---

[21] Kubik, *Theory of African Music, Volume II*, 112. See also Wegner, "Cognitive Aspects of Amadinda Xylophone Music from Buganda."

[22] The 'prediction error minimization' theory (PEM) says that the brain continually seeks to minimize its prediction error—minimize the difference between its predictions about the sensory input and the actual sensory input. See Hohwy, *The Predictive Mind*.

elements is a fact that may have a neurological basis in the limbic system, that is, the rewarding sensation we receive when noting consistencies or correspondences that convert an alien experience into a familiar one.[23] This is a process that these researchers referred to as 'feature binding,' which they illustrate using the textbook example of an emergent image, the black-and-white splotched image of a dalmatian in a garden. Upon close inspection, the dalmatian, which can be found secreted within an otherwise inchoate terrain of black-and-white splotches, will eventually pop into perception.[24] Elsewhere, Carsten Höller's[25] oddly captivating 1998 film *Punktefilm*, featuring Swedish folk dancers making their movements in darkness with lights affixed to various points on their bodies, provides another striking example of the same tendency towards this 'feature binding.' With these and similar works, creating a satisfying sense of 'closure' in many cases affords us a kind of status as 'co-creators' completing a work-in-progress rather than as passive consumers of aesthetic experience: "when artists speak of composition, or grouping," Ramachandran and Hirstein claim, "they are probably unconsciously tapping into these very same principles."[26] Their example of a "nude hidden with a diaphanous veil," and how it strikes us as more appealing than one shown without such obscuring features, also seems to hint at this joy in the creative process of actively overcoming perceptual challenges.

As we have already touched upon, listening is as active—and perhaps as *compositional* an act, if not more—as the other processes more traditionally understood as composition in the artistic sense: when we are listening there are a number of acoustic dimensions that we use in perception, as cues to group together the components of sound textures. Stream segregation theory has shown that whether we listen to our everyday sound environment or to music, we use many aspects of timbre to organize the musical surface into streams. The well-known Gestalt laws of organization largely guide the way we carry out such perceptual grouping, in a way analogous to how our sense of sight functions. Over time, we train our eyes to understand what features are unique to what object or class of objects, and we have aural faculties that perform roughly the same task for sonic information. For example, groupings dependent on similarity are common between the two senses (e.g., similarity in timbre and pitch for audition, and shape qualities for vision), as well as simplicity ('prototypical' objects are grouped in the same way as sounds with simple harmonic ratios) and proximity (elements 'close in space' for both audition and vision). When sonic elements that are perceptually similar are grouped together they will tend to form streams, as coherent *Gestalts*, whereas sounds that are perceptually

---

[23] Ramachandran and Hirstein, "The Science of Art."
[24] This is a classic example of an image with emergent properties which highlights the Gestalt school's idea of perception being holistic. R. C. James's Dalmatian photo made famous as such by inclusion in Gregory, *Visual Perception*, 14.
[25] Carsten Höller (1961–) is a German-born multidisciplinary artist known for ambitious large-scale projects and installations (many including live animals as a thematic element).
[26] Ramachandran and Hirstein, "The Science of Art."

dissimilar, under the same conditions, will tend to perceptually separate.[27] In music, many different factors or parameters, acoustic and perceptual, influence the stream segregation processes and the relationships between the horizontal (sequential) mode of sonic organization discussed above in the section on persistence, and the vertical (simultaneous) mode introduced in the section on layering.

In the case of my *Harp Phantoms* pieces, pitch and amplitude similarities or differences have an obvious role to play in the auditory stream formation, but, equally, timbral acoustic properties such as harmonic spectra, formants, attack and decay transients, due to the particular preparation of the instrument, as well as the repetitive nature of the material, will also affect the formation of auditory streams. In these pieces, there is a constantly changing formation of auditory streams and sub-streams, due to shifting shared borders between component patterns and subtle changes in timbral organization. A sound-producing action is repeated and built up into a stable looping ground pattern. After some time, factors such as subtle finger playing pressure changes can induce the once coherent and stable auditory stream (i.e., sonic results of action) to split into sub-streams. This splitting of a coherent stream into sub-streams can have a dramatic perceptual effect on the material, such as an alteration of the pattern rhythms or a sudden generation of independent noisy artifacts. Such noisy artifacts from the looping streaming structures are the predominant cause of the patterns we hear as sonic phantoms. Noisy timbral elements generated from activating the prepared strings of the harp combine and recombine with other sound elements to produce new timbral streams, which are both reinforced by the player and exaggerated with the amplification of the piezo-electric microphones.

With many of the harp patterns, if they are performed or replayed at slower speeds, we experience them as if they belong together to one group or Gestalt. They form a kind of coherent melody. At faster rates of play there is a point where this coherent pattern stream appears to splinter into multiple independent sub-patterns. Instead of hearing one stream of sound, we now hear a number of smaller individual streams. Due to the repetition of the material and the faster speeds, our perceptual apparatus spontaneously reassesses the input; we reorganize what we hear into new streams based on what our perception considers to be meaningful.

In general, the streaming of sounds will come into effect when rates of play reach around eight or more sound events per second, although this threshold will vary, depending on the type of material in question. At super speeds (fifteen sound events per second and higher), streaming processes break down and we can no longer process these independent streams anymore as separate distinct sound patterns, but instead

---

[27] Bregman's research into streaming also shows that we can distinguish between schematic and primitive processes in segregation. Schema-based processes are higher-level processes that depend upon knowledge-driven processes, whereas primitive modes of listening are pre-attentive and activate without the mediation of memory and learning. Primitive streaming in perception can be responsible for influencing the segregation of patterns based on factors including frequency separation, dynamic contrast, spacial separation, and timbral organization. See Bregman and Rudnicky, "Auditory Segregation"; Deutsch, "Grouping Mechanisms in Music"; Clarke, "Rhythm and Timing in Music"; DeWeese and Zador, "Neural Gallops Across Auditory Streams."

hear the groups as granular 'texture,' thus returning to the coherent auditory image. As part of this dynamic process, the Gestalt laws of similarity and proximity influence our perceptual grouping of sounds into streams such that sounds closest to each other in pitch or timbre will tend to stick together. As one would expect, the pitch distance between component sounds, that is, the interval, will have an effect upon the pitch and the timbral streaming of sounds.

My own experimentation (which concurs with research findings into stream segregation and formation) suggests that, when working with interlocking patterns, the most effective combined range of pitch intervals for efficient streaming will usually not exceed an octave. When conditions of speed and regularity are met, this will force an entire Gestalt sequence to split into sub-streams. These sub-streams can be experienced as prominent phantom patterns that seem to emerge magically from the interconnections between the sound elements.

In the scene #8 of *Harp Phantoms* the player uses an alternation of the thumb and first finger of each hand, successively looping each pattern to contribute to the overall pattern. In the scenes #6, #7, #9, and #10 I have divided the composed patterns, to be played alternating between both hands. The alternating hands should play the patterns, even and regular, equally spaced in time; despite the regularity, however, the interleaving of the two parts creates high and low perceptual streams that give rise to surprisingly irregular rhythms. The combined sequence is too fast for the ear to follow note by note and so our perceptual circuitry has to regroup the material to make the best fit, forming several different melodic and rhythmic patterns. The sonic phantoms that we hear are not played as such by the player but are instead a result of our pattern-forming perceptual mechanisms that attempt to group the incoming sounds into the most meaningful patterns. The splitting of sounds into independent streams provides potential for sonic phantoms to appear and disappear. Some of these patterns can serve to stimulate suggestions of words or phrases or suggestion of other sonic forms. Most of all, the oscillating patterns function as sonic blotscapes that suggest and inspire ideas for development of the patterns and the forms they might take. Intense listening of the repeating blotscape patterns will eventually 'animate' them, as if they were 'speaking out' or musicalizing themselves[28] (one example of a suggested verbal meaningful phrase I can hear in sections of the *Harp Phantoms* pieces is: 'I hit the wall').

A potent side effect of the streaming process is the possibility to generate irregular rhythmic and temporal distortion from regular patterned material. When tempo conditions are favorable for streaming, individual isolated elements will restructure themselves from the context of their host patterns to form new perceptual groupings. Consequently, the rhythmic structure of the pattern appears to have changed when, in actual fact, there has been no acoustic change of rhythmic structure. The streaming

---

[28] In relation to musical patterns in African culture played by lone players (Ennanga, Zither, lamellophone), Kubik writes about how the musician ends up 'reading' messages into the ever-changing oscillating patterns. 'Inherent patterns create a complex illusion of conversational polyphony, since many of these patterns are verbalized; they say something to the solitary performer and to audiences... quite often oscillating patterns also give composers new textual ideas'; Kubik, *Theory of African Music, Volume II*, 139.

effect causes a curious distortion of the temporal relationships between the elements and thus of our temporal experience of the music, including melodic and rhythmic patterns that can be entirely transformed upon making a certain change that affects their perceptual grouping. The notes of an interlocking pattern may be organized in a particular identifiable sequence in time—like beads in a necklace—but at fast rates, due to the interlocked structure, we perceptually group the elements that seem to make the best fit together. In scene #9 of *Harp Phantoms* two interleaving, isochronous, three-note patterns are played reaching the threshold speed, to give rise to new perceptual groupings. The former regular sequence transforms into multiple high and low streams based on pitch proximity and interpreted as irregular rhythms. What was initially an integrated rhythmic pattern constrained within one coherent stream seems to spontaneously split between two streams, and is therefore experienced as a change in rhythm. The regular rhythmic patterns are transformed into irregular ones due to the streaming process, and the perceived temporal order of the sequence of sounds is difficult to determine due to the ambiguity of the streaming process.

These spontaneous perceptual alterations are essentially what give rise to the sonic phantoms. In all the phantom pieces, in different ways, I manipulate the timbral stream-forming process to alter pitch and temporal patterns and so bring into audible focus new auditory streams that were previously embedded and undetectable in the main texture.

3.7 Sonic Figure and Ground

Auditory streaming is essentially, within the auditory domain, the counterpart of the figure-and-ground reversal images. Like the reversal bistable images, interlocking sonic structures that produce the streaming phenomenon also share common borders. In the bistable images, one can either attend to the figure or the ground, but not both at the same time. Similarly, when presented with an equivalent bistable situation that involves auditory streaming, listeners can only pay attention to one sonic stream at a time. This logic follows the Gestalt rule of 'belongingness' and is also described as 'mutual exclusion.' When a sound is incorporated into one stream, in general it cannot simultaneously become part of another auditory stream.[29] The particular stream that comes to occupy the foreground of attention would be the figure and the stream that we are *not* attending to would be the ground. It is the blurring of the boundaries in the foreground–background relationships that makes this perceptive situation bistable. If two patterns with similar timbral structures are interlocked and layered, as in the case of simultaneous harp sounding patterns, then both can alternatively be perceived as figure and ground at different moments.

Creating bistable systems with their resulting ambiguous patterns, which cause us to attend spontaneously to and alternate between figure and ground, not only offers

---

[29] Bregman and Rudnicky, "Auditory Segregation"; and Thomassen and Bendixen, "Subjective Perceptual Organization of a Complex Auditory Scene."

to the listener the interest of changing perspectives, but also provides the potential for generating sonic phantoms. In general, the more bistable or multistable the system is, the more ambiguous, unstable and unpredictable the figure-and-ground relationships will be. Ambiguous sonic scenes require a more active effort from the perceptual system to decode and we can play with this for creative results. We need to focus with intent in order to detect patterns before they emerge in our awareness. From my perspective, this effort may contribute to the aesthetic pleasure gained from the enjoyment of listening and ascertaining the ambiguity in a musical scene: those moments when clear patterns seem to suddenly reveal themselves.

In my phantom pieces, I work to compose many polysemous, 'open' auditory images; sonic blotscapes to be enjoyed without a single central focus and with the potential to be listened to from many alternating perspectives. Spontaneous patterns seem to randomly appear into awareness and then disappear into the sonic ground tapestry, perhaps never to reappear again. This capacity for different patterns to spontaneously rise to the foreground upon repeated listenings contributes to a perceptually unstable, and thus ambiguous, listening experience.

In general, subtle changes are better detected within more stable and relatively unchanging states; when everything is in a constant state of change, it is difficult to detect any distinct forms. A continuity of ground pattern established over time provides a stable constant background from which phantom patterns can emerge from and dissolve into awareness. For subtle timbral changes to be perceived and foregrounded, most other parameters should remain constant. In general, this means that changes made to patterns and their configurations should be steadily and slowly gradual. Making minute changes to one parameter at a time, whilst keeping all others constant, emphasizes and highlights the subtle timbral streaming effects and transformations resulting from that particular change. Changing too many factors simultaneously destabilizes the solid background, which is necessary for playing with the figure-and-ground relationships.

3.8 Listening Modes and Perceptual Competition

As mentioned earlier, the process of listening itself, at its different levels, will influence the perceptual experience of ambiguous musical material. We have seen, via Bregman's concept of Auditory Scene Analysis and audio stream segregation, how the perceptual mechanisms of our hearing generate meaning and sense out of the constant audio input; everyday, direct, unfocused modes of listening are used to detect and extract patterns more or less innately and automatically, and are understood to be pre-attentive or pre-cognitive. On a more focused and cognitively active listening mode, like that of aesthetic listening, we adopt schema-based modes to actively concentrate and zoom into sound patterns and their details. With these types of modes, we internalize patterns and make mental schemas out of them. When listening to repetitive sound patterns our listening modes shift between more inattentive and more schema-based segregation processes. Sometimes it will be the innate stream segregation processes that will guide and influence our perception, and sometimes we use our experience and repository of previously saved sonic schemas to consciously attend to certain groupings.

The key point to consider is that the degree of active concentration or *level of listening*, which activates the various streaming processes, will affect and influence how we perceive specific patterns, how we group sounds, and whether we audibly interpret them as motifs, melodies, textures, and so on. The attentional focus of the listener will often contribute to *how* the resulting streams are determined: by pitch streaming, by timbral streaming, by amplitude streaming, and so on. With ambiguous patterns, for example, pitch can compete for streaming with spatial location (e.g., panning considerations, musician placement). We may perceptually group the pattern sounds by how close the pitches seem to be to one another as opposed to grouping them by spatial location. In such a case, pitch proximity is dominant over spatial separation and those patterns grouped by pitch proximity or similarity will be foregrounded in our attention.

Being aware of and consciously setting up a situation of composing this kind of competition between the perceptual grouping cues can enhance the streaming effect and so contribute to the ambiguous outcome of the musical scene. Many of the sonic phantom streams that emerge from the interlocking patterns in the *Harp Phantoms* pieces do so as a result of competition between grouping cues such as pitch proximity or timbral similarity between sounds. The Gestalt laws of proximity and similarity[30] reveal why there is such ambiguity as to whether we perceptually segregate the acoustic scene by pitch or timbre.

3.9 Accentuation of Harp Sonic Phantoms

As a particular strategy of compositional 'illumination,' I have used in *Harp Phantoms* (as I have generally done in many of my compositions) subtle alterations of dynamics, spatialization, and frequency equalization of sonic materials, to be able to deliberately accentuate or highlight the presence of certain embedded streams and patterns previously obscured from conscious perception, in and out of focus. The accentuation techniques that I employ involve manipulating certain parameters of a pattern so as to enhance and exaggerate, with the purpose of bringing them fleetingly to the foreground. Composing by means of accentuating the emergence of alternate embedded patterns is, in effect, playing directly and intentionally with sonic phantoms.

For example, if the amplitude of a couple of notes within a certain pattern being played in the harp is minimally increased (by means of increased plucking pressure on a string, for instance), whilst keeping the rhythm and pitch pattern constant, then the influence of this new dynamic change will force the pattern to be foregrounded from its base texture. When the base pattern is played on the harp over the threshold speed, these individual amplified components will segregate from the rest of the texture. If

[30] See Handel, *Listening*.

highlighted, these sub-patterns will be heard as independent distinct patterns; but if the highlighting is removed, returning the pattern to its original form, then the notes will perceptually reintegrate into the context of the basic pattern. In other parts, the extreme high pitch of the plucked sound, which is repeated at random intervals in the pattern, helps to encourage a clear segregation of the grouping of these high components of the sound. Occasionally, the repetition of a single micro-element of the harp sound will clearly surface as an individual figure or foreground stream, which appears to have almost no temporal relationship to the interwoven fabric created by its underlying sonic texture. In a section of scene #9 of *Harp Phantoms*, the consecutive alternating notes G and A may be heard as a continuously alternating pattern. The close pitch proximity of these note patterns is responsible for the notes' perceptually grouping together, and makes it possible to hear them as a single coherent stream. This perceptual grouping is clearly heard as an independent stream above the remaining parts, which are interwoven in a tight configuration.

Another powerful strategy for bringing the phantom patterns into clearer focus, which I have used within the *Harp Phantoms* series, is reinforcement via establishment of coincidental events and layers. A salient example of this is provided in *Harp Phantoms with Organ* (a variation piece I created for a Wave Field Synthesis sound system),[31] in which I introduce a series of highly sustained organ sounds to coincide with the six-note harp pattern at the point when it has been established. Due to the simultaneous polyphony of the harp and organ layers, the sonic phantoms may not be heard until they are brought into focus. It should also be mentioned that use of coincidental material can either aid or hinder the ability to clearly isolate the distinct sub-patterns, depending on the context. In numerous cases, such simultaneous or coincidental processes conspire to create a push/pull effect of opposing directional forces. The sonic patterns created by the harp, with increased, non-linear internal activity, run parallel to a gradual lengthening pattern composed of a series of linear organ sustained notes. The linear progression of the organ material confers a sense of forward propulsion, and this in turn contrasts and competes with the apparent motionlessness of the harp patterns. This provides the listener with the opportunity to split attention between the linear or horizontal process of the organ, and the more vertical, internally moving, non-linear patterns of the harp.

Another interesting effect here involves a kind of flattening of the dynamism that would typically be expected of performed music. That is to say, illuminating a pattern embedded within the pre-existing texture can provide a perception of stasis rather than one of forward motion. Strong vertical relationships bring with them the potential for reorienting perception towards the more subtle details secreted within the sonic tapestry, in the process disrupting the customary impression of a horizontal or linear flow of time. Repetitive patterns typified by a tight interlocking of parts can counteract linear movement, which could result from timbral change or dynamic variation. All told, this work relies upon harnessing the potential of both linear and non-linear processes, which are arrived at through a process of trial and error. While playing

---

[31] www.gameoflife.nl

with the dynamics of these different temporal forces, I try to achieve illuminating contrasts between patterns that are perceptually ambiguous with those that are clear and distinct. This is another in the repertoire of techniques that can establish a stable ground from which the sonic phantoms can unexpectedly and mysteriously appear.

3.10 A Second Life of *Harp Phantoms* in the Studio

After a period of germination, consisting of a series of workshops, I came to identify a number of techniques and processes that could then be documented and scored for live performance. Sometime later I decided to make studio recordings of these actions and techniques, whilst they were fresh for me, and still fresh in the mind of the player as well. In the recording studio, we exhaustively documented the diverse actions and patterns that resulted from the initial experimentation period, a process that called upon a variety of different microphone configurations (for example, piezo or contact microphones placed directly upon the harp and then run through a custom-built Owen Drumm piezo mic mixer, or DPA omnidirectional mics placed inside the f-holes of the instrument). I have subsequently used this recorded material as a sort of sound bank of raw elements for making many new pieces, recycling these kernels of sonic information in a multitude of combinations and using them to form layers of patterns with either a dominantly synchronous or asynchronous character. At this point, I carefully considered the combinations to be integrated into future live performances, where they would be able to develop yet further once released into a performance environment loaded with additional vectors of unpredictability.

During the composition of the pieces *Harp Phantoms* #3, #6, #7, and #9 these techniques were borne in mind as a means to increase complexity of the original material, and to explore a more intricate polyphony than what might be possible with digital audio software and hardware. These studio pieces are composed using a process of stratification that involves simultaneous layers of sound; a vertical construction process that relies upon an accretion of simple component patterns. Layers are stacked and offset against each other to create densely detailed textures, and the subsequent repetition of both the layers and their offsets creates a vertical organization wherein symmetry and alignment of layers affect the resultant complexity of the overall audible pattern.

Typically, I carried out this studio work in two consecutive stages. The first one involves setting up two or more patterns to play together in synchrony, each one of them aligned and panned to center position. To establish this as a stable ground pattern is the first step to be achieved in the sonic phantom-production process, and this requires enough exposure over a certain duration of time, this being dependent on the length of the pattern itself. In the second stage, which occurs once these 'ground' patterns are stable enough to be internalized, subtle micro-shifts of a second pattern can be made; layers can be shifted or nudged to the degree of a few frames per second or an even higher resolution (smaller units of time) in digital environments, which allows for these micro-shifts to take place. A prototypical example of this two-stage process would start with an action or pattern 'A' composed and recorded in the studio, and repeated continuously in one audio track (A-t1). This would be the 'ground' stable pattern. An identical copy of this pattern is made digitally and then played in unison with the original on a second track (A-t2). A-t2 is nudged incrementally by hand, and using digital samples as the unit of measurement, as A-t1 loops continuously and uniformly. The process of offsetting track 2 against track 1 by hand is then recorded in real time on the computer. The first incremental nudgings of the phased second layer, A-t2, result in slow and subtle changes of resonance and timbre.

A point is eventually reached whereby we hear what are apparently sudden and rather unexpected shifts in rhythmic activity. Despite the fact that these movements have been consistently gradual over time, and thus could be said to deny the possibility of unanticipated revelations, there quite suddenly appear to be spontaneous moments of instability and irregularity. As the patterns continue to repeat and phase against each other, these irregular interstitial moments eventually gather stability again.

These unstable transitional or liminal zones that become evident during the nudging process are themselves attributable to perceptual regrouping processes; the auditory streaming that we have already explored. This fragmentation stage can feel chaotic before it ultimately resolves in a new state of stable and coherent order. New patterns are then established, and then again begin to fragment with the phased asynchronous shifting of the cycles. The experience of transitioning from order to disorder, or vice versa, promotes the appearance of sonic phantoms: sometimes gradually, sometimes with an unexpected immediacy. As they fade in and out of our conscious attention, our perceptual mechanisms are forced to regroup each time to make the best sense of the auditory ambiguity, guided towards the detection of new patterns and configurations with the aid of the primitive and schematic stream segregation processes.

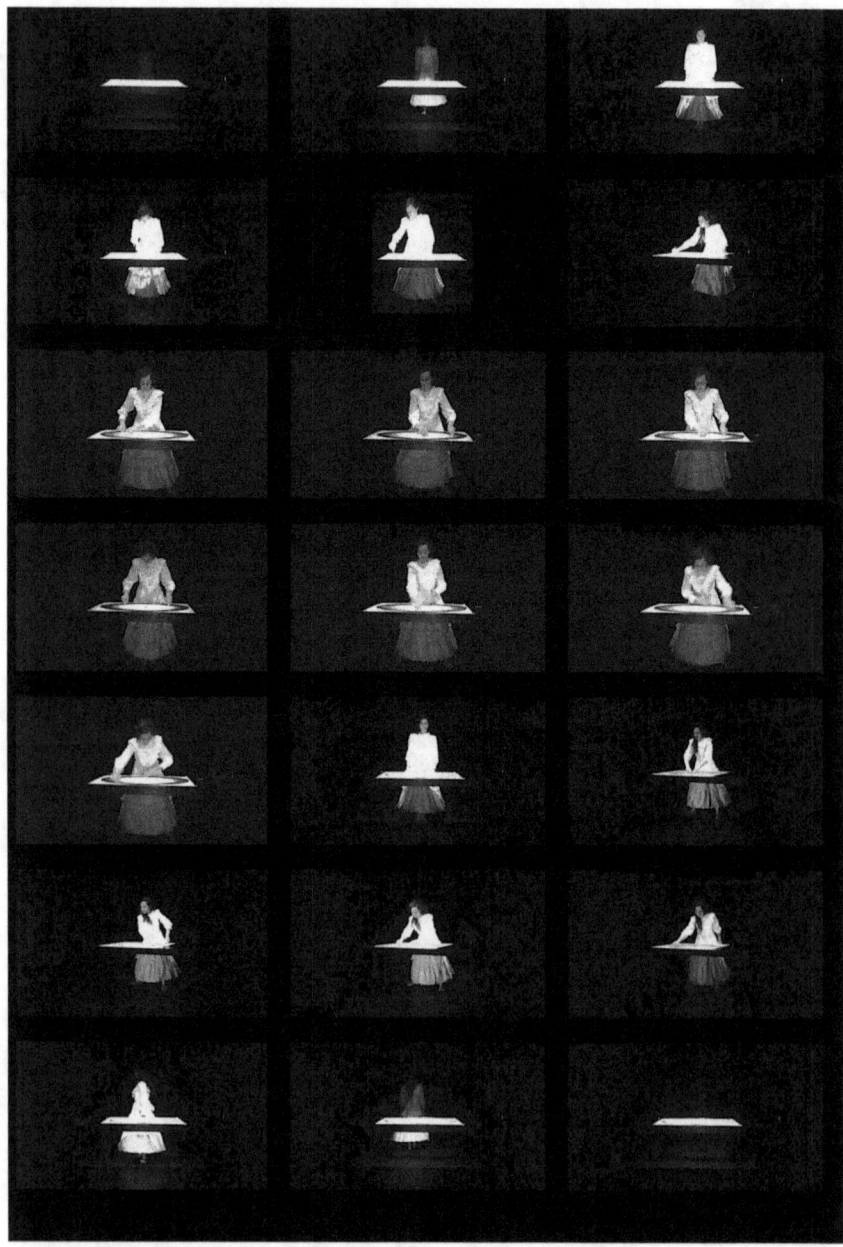

Figure 4.1 *Drawing Phantoms* séance. Sonic phantom composition by Barbara Ellison, performed by Nathalie Smoor. Photographs by Barbara Ellison, 2013.

4

Phantasma Materialis: The Realm of the Object

Compositional Series: *Drawing Phantoms*

4.1 Initial Material Explorations (for a Literal 'Phono-Graphic' Performance)

In this chapter I introduce a body of work developed over the course of several years, collectively entitled *Drawing Phantoms*, to use as the framework for discussing characteristic aspects of my compositional practice, and further elucidating how ideas of trance induction and the ritual process relate to the production of sonic phantoms.

As with other examples of my work that have been discussed to this point, this material has involved a primary phase of experimentation followed by a performance phase and the documentation thereof. These pieces essentially explore the production of perceptual sonic phantoms through the ritualization of a limited set of actions, a feature that is held in common between them and all the other phantom-production techniques I have experimented with. They play with ritual form, to varying different degrees, and give rise to phantasmatic sonic structures, which emerge through a hypnotic repetitious exploration of a simple sound-producing action.

To induce this phantasmatic state, the pieces in the compositional series *Drawing Phantoms* depend mainly upon using drawing tools to mark and transform circular trajectories on amplified surfaces, which lead to the generation of phantom voice-like sounds. These voice-like sounds are reminiscent of effects experienced when listening to EVP recordings ('Electronic Voice Phenomena'; a field of paranormal audio research described in detail below). This strange psychoacoustic effect owes itself to the slippery shape-shifting nature of the audible sonic image as interpreted by the brain, coupled with the trance-inducing effect of the repetitive action (and keep in mind again that the 'interpretation' phase here can go hand in hand with a 'generating' phase, given what we now know about how predictive perceptual models work). The act of drawing abstract reiterative trajectories could be described as 'phono-graphic,' in a sense that deviates from the commonly accepted meaning of that term but that is even more literal (i.e., combining auditory and visible elements) in the performance of these pieces: the 'drawing of sound' or the 'sounding of drawings,' which I use to create a particular kind of trance-inducing, audio-psychological experience.

Readers can listen to several pieces of this compositional series in the website www.sonicphantoms.com

In addition to the psychological alterations that can affect the performer, this systematic drawing activity can also be quite a physically demanding process, in that it might require considerable stamina when performed over extended durations of time. I have occasionally described my process as one of 'drawing EVP,' though I will elaborate later, however, on why it would be much more fitting to describe it by a term that is less historically loaded or suggestive by comparison: that is 'OVP,' or 'Object Voice Phenomena,' which is Francisco López's proposed term for my attempted revelation of a much broader and deeper phantasmatic phenomenon that does not completely overlap with what is created via electronic means. In essence, this is a series of creative actions centered on exploiting the full potential of our brains to hear semantic form or content in random structures. It is therefore an exploration of auditory apophenia. Whether we are speaking of the EVP tradition or of my own OVP practice adapted from that, the psychological, highly personalized mechanism of 'projection' plays an important role in the hearing of voices: it seems that there as many possible interpretations of the ambiguous audio material as there are distinct listeners.

I can trace back the inspiration for my drawing pieces to a clear starting point at the beginning of these compositional efforts many years ago, when I first began to explore the use of mechanical repetition for the purpose of 'exciting' objects, so as to create sonic looping structures. On one occasion, I chanced upon a set of square concrete blocks with their own set of charming inherent resonances, and began to use them as sounding boards. I found myself completely entranced by the sounds they made, as I used a small piece of chipped stone as kind of stylus for etching noisy, coarse circles and infinity symbols continuously on the surface. The mechanically repetitive drawing process, and the resulting array of rough and noisy sounds, brought about an unequivocally hypnotic sensation. Shortly after this first round of sonic exploration, I then experimented with amplifying the blocks with contact microphones to reinforce and magnify the more subtle sound patterns produced. I was really amazed at the range and variety of timbral sonic possibilities that arose from such an apparently limited starting material.[1]

From the outset, this was a very intuitive and elegantly simple, playful process, from which a full complement of rich and complex sonic phantoms were generated over time, and given an enhanced degree of detail through my awareness of them. This sparked my interest in the generation of sonic phantoms from the repetition of simple processes with similarly uncomplicated (read: 'everyday' or 'close at hand') materials. Over time, I have explored how such varied actions and gestures, none of them particularly difficult for non-specialists to intuit and reproduce, can be applied to a broad range of materials and can be ritualized and exploited for their sonic phantom generation potential.

---

[1] These initial explorations and subsequent evening-long performance of my *Phase portrait* project was presented as #5 of De Player Compost Series, at "het Gemaal" Rotterdam, 2008.

4.2 *The Drawing Room*

The Drawing Room was the initial phase of the *Drawing Phantoms* project and first realized as a collective performative art installation in the main hall of an old school,[2] where wooden tables from one of the disused classrooms were put into service as resonant surface structures, then amplified with attached piezo-electric contact microphones. These tables were used as drawing surfaces for an intimate ritual, a daily performance of sonic drawings for both children and adults as players (four to six participants at a time), sonically projected through a quadraphonic sound system. The event's focus was to use the hypnotic power of repetitively drawing shapes in order to bring about a meditative and semi 'trance-like' state. The shapes selected were thus intentionally simple so as to better lend themselves to repetition, regularity, and smooth transitions between them (e.g., circles, ovals, and infinity signs). Upon completion of the ritual performance, the drawing trace of this process leaves behind beautiful and intriguing colorful abstract trajectories.

I took various steps to ensure a ritualistic atmosphere before participants entered the space, and also took care to minimize sonic distractions from the sounds associated with the drawing process. Each performance group comprised of twelve players, divided into four groups of three. Each group were given color-coded aprons to identify which of the four performance tables they would work from. When the participants were ready, I explained and demonstrated the process (without amplification, so as not to spoil the sonic surprise) and careful but simple instructions were given to reinforce the instructions in the score, printed copies of which the participants had seen posted before the performance. Lights were dimmed and silent attentional focus was directed to the blank paper of the four tables. Upon a cue, the sand timers (a strategic means for timing the drawing process without creating additional audible distractions) were turned and the drawing ritual began.

This piece was arranged and composed to maximize the engagement of visual, aural, and somatic faculties that are aggregated together and recognized as 'attentiveness'. To this end, the contact microphone amplification of several simultaneous repetitive actions of this kind gave rise to dense, complex, and variegated sonic textures in the space; in most instances, with a mesmerizing effect accompanying them. Some of the participants reported that within this atmosphere of calm and concentration the ritual drawing of the sounding shapes had a 'magical' experiential quality to it (one participant spoke of her experience of each shape being a magical symbol, which slowly released its power over the drawing period), itself heightened by the socio-cognitive effects of the performance being a public one.

There is a sense of expectation and of mystery when the participants are aware in advance that deeply absorbed and trance-like feelings may be experienced after some time. This expectation led to extreme concentration, seriousness (in the best possible sense of the term), and a perceptible commitment from the participants during the drawing process. Many of them described sensations of 'entering into another

---

[2] *The Drawing Room* at Zinder festival, Stella Theatre, The Hague, 2012.

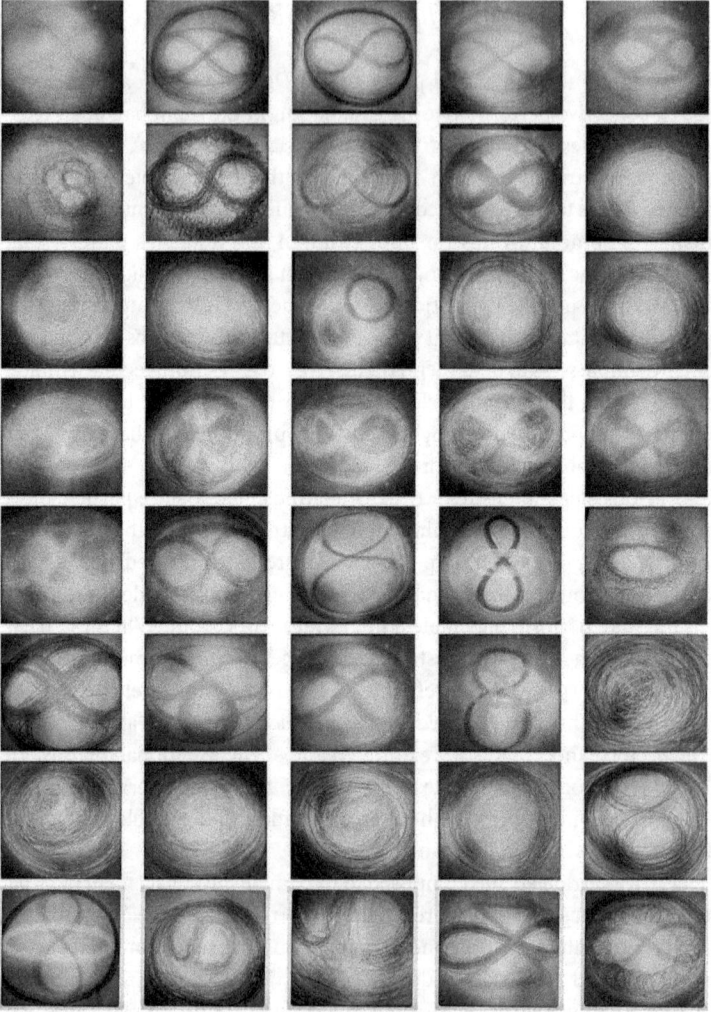

Figure 4.2 A selection of prototypical 'traces' (final graphic results) of the drawing process from the collective performance *The Drawing Room*. Photographs by Barbara Ellison, 2012.

world,' of feeling 'spellbound' in the process. Feelings reported in general were of the kind of deep meditation, trance, relaxation, calm, feeling energized, and—most significantly—of hearing 'voices' that, without any previous admonition, arose from the noise textures produced by the drawing action. With the added suggestion that it was a possibility, voices, melodies, and rhythms (different from the apparent ones produced by the hand motions) were unanimously 'heard' from the granular sounds of the collective drawings. Playfully suggesting that the drawn symbols had a magical character served to induce an excitement of the apophenic imagination. There was

Figure 4.3 *The Drawing Room* collective performance. Performance space: set up with school wooden tables, contact microphones, sand timers, wax crayons, paper, and metal clips; volunteer children and adult performers; and drawing final trace results of some of the performances. Stella Theatre, The Hague, The Netherlands. Photographs by Barbara Ellison, 2012.

intense focus and concentration to work these 'magic spells' with the drawn voice-producing symbols. The participants reinforced each other as they collectively moved in and out of synchrony in their drawing movements. Many also spoke of the sensation of uncontrollable entraining (they used terms like 'syncing up') to the drawing speeds of the drawing participant beside them.

4.3 Transcendent 'Boundary Loss'

In *Drawing Phantoms*, as with the previously presented compositional series *Harp Phantoms*, the repetition and the ritualization of the sound-producing action itself plays a major part in the induction of trance-like experiences. Repetition in this context is naturally a key strategy used to induce states of deep absorption in both the listening and the performing process. In general, the powerful effect of enacting repeated movements (for example, in dance, sport, or ritualized collectives such as those of marching drills) is widely known and deliberately used for different purposes of this nature. This is elegantly described and analyzed in great detail by William McNeill in his book *Keeping Together in Time*,[3] which discusses the phenomenon of 'moving together' by repeated and coordinated movements, and also makes a case for this phenomenon remaining fairly constant throughout human history. In accordance with the observations of McNeill and others,[4] and of course also more generally with virtually every single tradition of ritual, it is my experience that the repetition of actions or movements over long periods of time can produce states of dissociation, which can eventually lead to what we would normally consider 'trance.' It should probably go without saying that the same holds true for the sonic environments created during these activities.

McNeill also describes this 'moving together' phenomenon as 'muscular bonding,' which can lead to 'boundary loss.' Writing more specifically about contemporary dance, he describes the feeling experienced of 'losing oneself' as "... the submergence of self into the flows."[5] As we know, absorption and elation abound in the realm of dance:

> As the dancer loses himself in the dance, as he becomes absorbed in the unified community, he reaches a state of elation in which he feels himself filled with energy or force immediately beyond his ordinary state, and so finds himself able to perform prodigies of exertion.[6]

The immersion of the dancer, as executor of ritualized repetitive actions, insistently points to a blurring or loss of boundaries, or self-awareness, where consciousness is altered as the music or movement takes hold.[7]

In *The Drawing Room* piece, the activity of moving together in and out of synchrony with others provides an experience that fundamentally differs from that of performing solo. Most significantly, the practice of sonic drawing with others eradicates a certain performance pressure, in terms of keeping the drawing movements smooth, active, and continuous. It is a collectively emerging sonic result, as each participant affects

---

[3] McNeill, *Keeping Together in Time*.
[4] Bücher, *Arbeit und Rhythmus*; McNeill, *Keeping Together in Time*; Gioia, *Work Songs*. See also Brown and Volgsten, *Music and Manipulation*.
[5] McNeill, *Keeping Together in Time*, 8.
[6] Radcliffe-Brown, *The Andaman Islanders*, 257.
[7] Burt, *The Judson Dance Theatre Performative Traces*.

each other's pace and rhythm. This leads naturally to entrainment, a condition that chaos theorist Steven Strogatz describes as the "process of external synchronization, of falling in step with the outside world."[8] This synchronization with other performers arguably achieves its effectiveness from its status as a microcosm of other biological and neurological processes governed by entrainment.[9]

It is interesting to realize how the sonic layering process works here: with the uncontrollable tendency to entrain with the other performers, there is often the unconscious effort to synchronize, an effort that is sometimes successful, but mostly very changeable and unstable, which leads to the constant temporal displacement of layers, moving in and out of synchrony. This contributes to the richness and complexity of the layering, as well as to the subtle differences in timbral quality and granularity between each participant's sound drawings. This produces a kind of chorusing effect, which in turn gives extra texture and granularity to the overall sound.

In the *Drawing Phantoms* pieces the idea of the open perceptual mind is, from my perspective, crucial to the process, as openness and expectation enable the ability to intensely focus and concentrate on the action and sound; and, in doing so, exclude all surrounding factors from perceptual awareness. This narrowing of the perceptual focus enables the space for the imagination to take flight. With an open mind and commitment to the process, prolonged exposure to repetitive movements, or sounds, or images, serves to activate the feeling of boundary loss and absorption.

4.4 Automatic Drawing—Ghosts and Dissociation

In the liminal world of the Victorian spiritualists, 'automatic drawing' techniques were often used by mediums as a way to induce an internal trance or 'channel a spirit' (said technique has also been adopted by groups as diverse in their origins and intentions as the Theosophical Society and the original Surrealist artists). Spiritualist mediums would appear to fall into a trance during a séance and produce drawings under these heightened conditions. Claims were then made that ghosts or spirits were driving the hands and were the actual sources of the images. Through reflection on my own drawing rituals, I was naturally drawn to these practices and felt attracted to the intertwined similarities between both of them: my use of drawing as a way to induce

---

[8] Strogatz, *Sync*.
[9] Entrainment is a process where at least two or more autonomous rhythmic processes interact with each other so that they eventually synchronize or 'lock in' to a common phase and/or periodicity. It is defined therefore as a synchronization of two or more rhythmic processes and was first discovered by Dutch scientist Christian Huygens in 1665 with his experiment of 'the sympathy of clocks' relating to how a room of pendulum clocks after one day have all synchronized. Examples of entrainment include fireflies illuminating in synchrony, humans adjusting their walking patterns to match each other, brainwaves 'entraining' to certain frequencies, etc. Auditory entrainment is also known as 'auditory driving.' See Clayton, Sager, and Will, "In Time with the Music"; Clayton, "What Is Entrainment?"; Neher, "Auditory Driving Observed with Scalp Electrodes in Normal Subjects"; and Phillips-Silver, Aktipis, and Bryant, "The Ecology of Entrainment."

the phantasmatic and the use of those automated drawing techniques as a way to access intensely heightened states of focus and creativity.

Automated drawing as a technique had widespread use during the nineteenth century—and is still in use today in certain spiritualist and artistic circles—specifically to achieve an 'empty' or focused state of mind that could be sustained for a prolonged period of time (the entire duration of the automatic drawing exercise).[10] The term itself is somewhat misleading, as there is in fact nothing wholly 'automatic' about the process (it is also important to, *pace* Pierre Janet, distinguish the 'motor automatism' of practices such as automatic writing and drawing from the less voluntary 'sensory automatism' represented by hallucination). It is nonetheless mainly used to describe the process whereby images and forms can seem to appear without conscious effort or design, as a result of an intense subconscious drawing session.

The English artist and occultist Austin Osman Spare, perhaps one of the more interesting figures in the recent history of this technique, arose in roughly the same era, and is a valuable subject of study not merely for his forming the 'missing link' between Victorian spiritualism and key ideas of the Surrealist movement (which he prefigured by at least a decade), but also for showing the connectedness of automatic drawing to trance states. Artist and critic Joseph Nechvatal,[11] in his panegyric on Spare, also argues for his relevance to yet more recent movements in contemporary thought and creativity: he acknowledges that Spare's "experiments with trance" foreshadowed "reflexive feedback loops— the basis of cybernetics,"[12] while also anticipating some of the related discourse of postmodernism.

Spare is known these days as much for his sidereal artwork as for his development of a highly personalized occult system based on the redemptive nature of chaos, and the possibility of perichoresis (a Christian theological concept relating to the indivisibility of the Holy Trinity, but also one utilized by fantasists like Arthur Machen[13] to refer to an interpenetration of dimensions).[14] As to the latter, Spare is in fact one of the 'patron saints,' along with Neils Bohr, of the 'chaos magick' school formed a little over a decade after his death. It is also worth noting, given our prior consideration of the trickster archetype, that Spare made liberal use of the satyr figure in his artwork: whether developed in a moment of trance-induced automatism or not, the continual

---

[10] Maclagan, *Line Let Loose*; Jones et al., *Drawing Surrealism*; MoMA, "The Collection | MoMA."
[11] Joseph Nechvatal (1951–) is a 'post-conceptual' French/American artist, composer, and author, whose running interests have included the transformative potential of noise and the hybrid theory of 'viractualism.'
[12] Nechvatal, *Towards an Immersive Intelligence*.
[13] Arthur Machen (1863–1947, born Arthur Llewllyn Jones) was a Welsh author significant to the development of supernatural fiction (see, e.g., his *Great God Pan*).
[14] This is relayed through characters in certain of Machen's stories, e.g., a character in his 1936 story 'N' who muses on how a sense of 'jamais vu' points towards overlapping of different universes or planes of existence:

> Has it ever been your fortune…to rise in the earliest dawning of a summer day, ere yet the radiant beams of the sun have done more than touch with light the domes and spires of the great city?…If this has been your lot, have you not observed that magic powers have apparently been at work?

identification with this avatar of the mythological trickster says much about Spare's self-perception of his mischievous, yet tutelary role vis-à-vis the public, as well as his unyielding campaign to alter perception in general.

Spare's enthusiasm for trance states was at the center of his creative practices, extended to such activities as the ominously named 'death posture' ritual:[15] an immersion in ego annihilation that sounds singularly terrifying when explained in Spare's ornately wrought prose, yet is not far removed in its extremity from other shamanic rites of antiquity. See, for example, the Celtic initiatory rite of wrapping oneself in the hide of a recently slaughtered bull until visitation by oracular entities, or, more notably for audio enthusiasts, the Tibetan Buddhist practice of *chöd*: a purification rite that partly involves playing music on a human thighbone trumpet within a charnel ground until deities and demons come to feast upon the ego of the adept. Spare's practice was predicated upon "clearing away the superficial so that we may recognize our identity with the immensity that lies watching from within,"[16] and as such it was really a more intense means of achieving results that could be brought about by automatic writing or drawing, that is, "[giving] his deep psyche a chance to describe itself on its own terms...without the interference of the ego."[17] The same end result was also at work in Spare's 'sigils,' glyphs utilized as wish-fulfillment devices once certain formulaic instructions had been followed (e.g., write one's statement of willful intent, remove the duplicate letters in the statement, and arrange the remaining letters in such a suitably unique device that would be "too abstract for symbols, and too complicated for letters").[18]

Borne on the wings of bold pronouncements such as "the poet to come will surmount the depressing idea of the irreparable divorce between action and dream,"[19] the much more commonly known automatic writing and drawing efforts of the Surrealists would follow soon enough (and it is strongly suggested that there was an "overlapping of these two graphic modes").[20] The oneirically inspired name of the Surrealist movement, first coined by Guillaume Apollinaire[21] in 1917, was more clearly defined by its outspoken ambassador André Breton[22] in the 1924 group manifesto, and ever after was introduced to the uninitiated as "a certain psychic automatism that

---

[15] First introduced in Spare's *Book of Pleasure*, this technique is described by Spare acolyte Stephen Mace as an "advanced technique for conjuring" involving tightening of the body and "breathing 'deeply and spasmodically.'" Through such action Spare insisted that the "wizard...forcing his ego to mock death...can 'stand back' to see the powers that energize his own and others' actions, and so learn how he can best work to carry out his will." See Mace, *Stealing the Fire from Heaven*.
[16] Ibid.
[17] Ibid.
[18] Fries, "Visual-Magick."
[19] Breton and Parinaud, *Conversations*.
[20] Nixon, "Dream Dust."
[21] Guillaume Apollinaire (1880–1918, born Wilhelm Albert Włodzimierz Apolinary Kostrowicki) was a French poet credited with bringing the genre-defining terms 'cubism,' 'orphism,' and 'surrealism' into common currency.
[22] André Breton (1896–1966) was a French poet and polemicist whose 1924 *Surrealist Manifesto* laid much of the groundwork for future efforts conducted under the banner of surrealism. "UbuWeb Papers: André Breton—Manifesto of Surrealism (1924)." http://www.ubu.com/papers/breton_surrealism_manifesto.html. Accessed January 2, 2020.

corresponds rather well to the dream state."[23] Though this was the proclamation of the man who many would later deride as an autocratic 'pope' known for excommunicating fellow Surrealists on a whim (a personality hinted at by Breton's qualifying statement, "I have never lost my conviction that nothing said or done is worthwhile outside that magic dictation"), it stands as well as any current definition of Surrealist motivations.

Having noted all this, focusing on Breton exclusively is unwise since numerous fellow travelers of his are worthy of further examination. One such is the French artist André Masson, who was also renowned for his automatic drawings. These drawings functioned for him as essentially equivalent to the previously discussed 'blotscapes', acting as a gateway to the subconscious achieved by the exploitation of chance effects and unexpected juxtapositions. Masson would start drawing with no plan or composition in mind, and, without conscious control, he would begin allowing the pen 'to guide' the process and speed of the movement. After some time, forms would be detected in the abstract markings: various forms (objects, bodies, animals, etc.) from his own internal image bank would emerge, seemingly unconscious to him, in the process. Sometimes he would reinforce these forms to illuminate them and bring them clearly to the surface; other times they would be left in that liminal state of ambiguity. Whatever one believes the source of the drawings to be, the goal would be to find a way to access the creative subconscious and, to use Marcel Duchamp's famous expression, to 'forget the hand.'

Meanwhile, despite his declared interest in and occasional defense of the work of Society for Psychical Research founder F. W. H. Myers[24] (to the extent of praising his opus *Human Personality and its Survival After Bodily Death* as "beautiful"),[25] Breton's cautious retreat from the more paranormal aspects of the automatic creative act differentiated him from the ambitions of a thoroughly convinced occultist like Spare while setting the stage for a more post-Freudian analysis of, and exposition of, the subconscious. In other words, the use of automatic writing and drawing techniques was intended to "transcribe, or trace, the never seen into the condition of audibility or visibility." This exposition of the technique's aims already strongly hints that automatic writing was not the only possible method towards attaining this condition, and the acknowledgment of 'audibility' does more than just crack the door open for future experiments with sonic art. When Breton goes on to invoke the trance state, it further cements the connection between these practices and the shamanic rites of ecstasy, which were, again, inconceivable without their audio component. In his own words:

> First and foremost, we must understand that the processes governing the practice of automatic writing and the study of outer manifestations of induced slumber were one and the same. In both cases, what we were trying to reach and explore was none other than so-called *trance-like states* [italics in the original].... What

---

[23] Gray, *The Immortalization Commission*.

[24] Frederick W. H. Myers (1843–1901), as the founder of the Society for Psychical Research, was one of the key intellectual contributors to the spiritualist movement.

[25] Myers and Smith, *Human Personality and its Survival of Bodily Death*.

fascinated us was the possibility they offered of escaping the constraints that weigh on supervised thought.[26]

Making good on Spare's pronouncement about the role of the "poet to come," Surrealism was in the beginning very much a poet's movement, with poetic recitation being perhaps the closest that the original configuration would come to any sort of incursion into sound or music. However, the application of psychic automatism to visual compositions and eventually film (e.g., the infamous collaborations with Luis Buñuel and Salvador Dalí) would arguably provide the inspiration for sonic efforts initiated well after the passing of the original Surrealist instigators. Despite their status as 'unofficial' Surrealists, many latter-day composers and amorphous musical collectives have capably applied the idea of psychic automatism to the post-industrial world. For example, the famed Surrealist technique of the *cadavre exquis* ('exquisite corpse'), involving the collective assembly of a complete text or image from individual parts whose exact details the other contributors are unaware of, was deployed in the 1940s by a group of composers including John Cage, Virgil Thomson,[27] and Lou Harrison[28]—at which time the French-Canadian artistic group called *Les Automatistes* was also using automated drawing techniques. Alvin Lucier has also encouraged the use of automatic writing techniques as a pedagogical tool ("in some of my composition classes, the first thing I do when I see that student composers are stuck for ideas is to suggest that they try automatic writing... it's a way to get unstuck when you can't work, it can give you beautiful and unexpected results").[29] In Japan, Group Ongaku— whose membership included the future inventor of the 'prepared compact disc,' Yasunao Tone, alongside Takehisa Kosugi[30] and Shuko Mizuno[31]—felt that their "improvisational performance could become a kind of automatic writing."[32] Tone argues for this on the basis that, though no text was involved in these improvisations, other contemporaneous media, such as the "drip painting of Jackson Pollock," shared the same methodology and intent.

The legacy of the automatic writing technique has also persisted beyond what could be termed the 'classical' period of avant-garde development into the 1980s, 1990s, and beyond: even when not obviously attempting to translate the technique into audio performance, numerous artists have paid tribute via album or track titles and other

---

[26] Breton and Parinaud, *Conversations*.
[27] Virgil Thomson (1896–1989) was one of the preeminent developers of American classical music, influencing the later work of artists like Leonard Bernstein and Paul Bowles.
[28] Lou Harrison (1917–2003) was an American classical composer conspicuous for his microtonal technique, use of non-Western instrumentation (specifically Javanese gamelan), and novel motifs known as 'melodicles.'
[29] Lucier, *Music 109*.
[30] Takehisa Kosugi (1938–2018) was a Japanese musician and composer whose key instrument was electronically treated, droning violin, and who participated in both the local Group Ongaku and international Fluxus movements.
[31] Shuko Mizuno (1934–) is a Japanese musician and composer whose works have extended from jazz ensemble work to chamber music and choral works.
[32] Tone, *Yasunao Tone*.

textual ephemera, such as Bernhard Günter's[33] track *Écriture Automatique* from the *Details Agrandis* album from 1998. Elsewhere, there is Nurse With Wound,[34] one of the more influential musical projects to self-identify as Surrealist and to create music that occasionally verges on the mesmeric (their *Sugar Fish Drink* CD compilation is subtitled *A Layman's Guide to Cod Surrealism*). Incidentally, their captivatingly strange piece 1986 *A Missing Sense* was a structural homage to another memorable piece of work deriving from this technique, Robert Ashley's[35] voice-driven piece *Automatic Writing*. Though not clearly deriving inspiration from written text, as the title might suggest, it is nonetheless an example worth including in this inventory and worth contrasting with the foregoing items. The 48-minute piece dating to 1979 features an alluringly distant electronic backdrop of Polymoog, organ and custom-built circuits, against which Ashley and co-vocalist Mimi Johnson perform a litany of 'involuntary speech' samples that feel uncomfortably intimate at times. Ashley was inspired to commit to this recorded exercise when understanding that

> I might have a mild form of Tourette's syndrome (characterized in my case only by purely involuntary speech) and I wondered, naturally, because the syndrome has to do with sound-making and because the manifestation of the syndrome seemed so much like a primitive form of composing—an urgency connected to the sound-making and the unavoidable feeling that I was trying to "get something right"—whether the syndrome was connected in some way to my obvious tendencies as a composer.[36]

Ashley also notably mentioned that the other compositional elements or 'characters' in this piece seemed "as unplanned, even 'uncontrolled' in their various ways as was the text of the involuntary speech." As divergent as his method may be from those methods of automatic writing (and not merely because writing itself is absent in his method), Ashley's work still aims at a type of undiluted vision that has long been the goal of this practice in all of its different iterations. The laconic murmurs, in many cases indistinct from the types of somniloquy or sleep talking, are yet another form of audio pareidolia for us to infuse with semantic value (the same also goes for Johnson's French-language whispers, and some of the vocalizations on Nurse with Wound's loose tribute to this piece).

Besides the illuminating and contextualizing relevance of the above cultural examples from the avant-garde and experimental arts communities, we should not forget that a great many of us have experienced and naturally delighted in intuitive

---

[33] Bernhard Günter (1957–) is a German-born musician and proprietor of the Trente Oiseaux record label; both his own efforts and those curated for that label contributed to the formation of the "lowercase" musical style typified by sparseness of audible events.

[34] Nurse With Wound are a UK-based music project whose large discography has been instrumental in bridging the gaps between a psychedelic aesthetic and the more concrete techniques of industrial music.

[35] Robert Ashley (1930–2014) is an American composer most known for his new forms of opera and multidisciplinary projects. He pioneered opera-for-television.

[36] Quoted in liner notes from *Automatic Writing*. Ashley, "Automatic Writing [1996, CD]."

forms of 'automatic drawing' in our childhood. Sadly, as far as the potential of this intuitive capacity goes, it is often the case that in later years we are taught to unlearn this spontaneity and to draw 'properly.' When the obvious and logical conventions about technique and control take hold, we tend to quickly lose this natural and spontaneous ability. Somehow we all develop the capacity to be highly competent 'automatic' creators, until many of us are led to believe that we cannot— 'really,' 'seriously'—draw.

I have my own personal technique of repetitive automated drawing, which functions, very much like the aforementioned examples, as an effective meditative method for accessing states of heightened creativity. Though it eventually developed into a performance, this is more of a daily ritual, mostly done in contexts that do not require public participation or acknowledgment, that is, alone as a meditative act. The

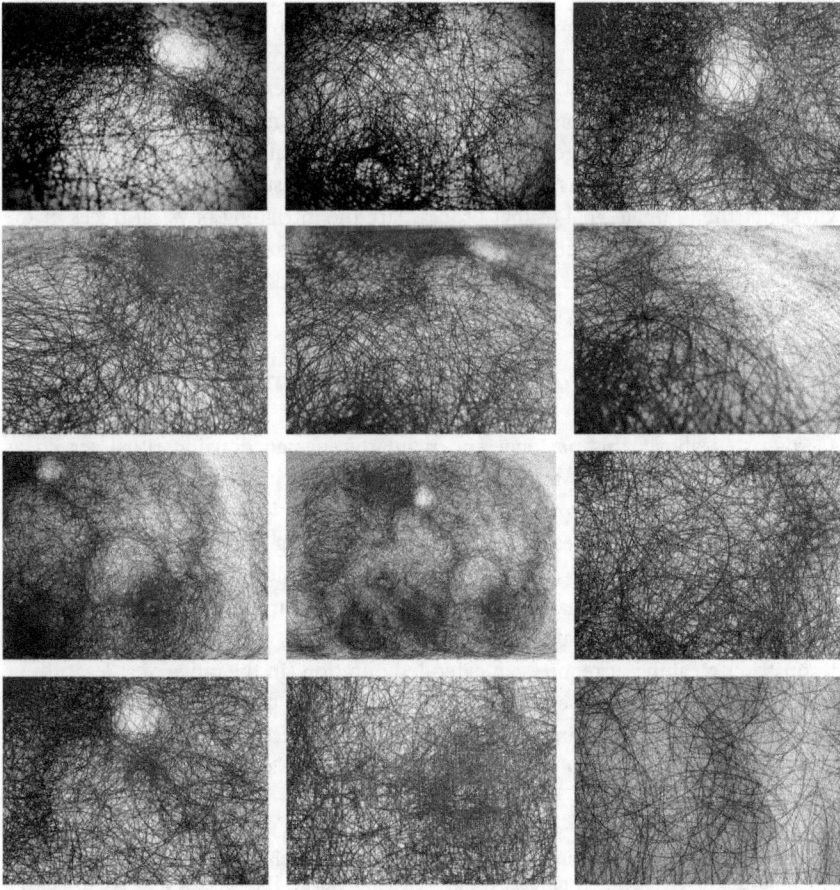

Figure 4.4 A few selected graphic results from some of my personal 'automated drawing' sessions. Photographs of A0 size. Original drawings by Barbara Ellison, 2013.

repetitive drawing process facilitates a way to forget the conscious self, to 'dissociate', and in so doing to bypass personal aesthetic preferences through encouraging immersion in the act itself.

As a premeditated and deliberate action (and taken very seriously and rigorously, despite what it might sound like to some), I have spent many hours drawing circles on all kinds of surfaces, during which time I often have the illusory feeling that an external force is driving the pencil. This uncanny phantasmatic sensation can be explained by what is known as the 'ideomotor effect'[37]—a movement caused by the idea of its realization—but it nonetheless is felt as being very real and even supernatural. The main objective is, in any case, to engage fully and completely in the process of circle drawing, without an end result in mind, and to 'inhabit' the movement as the pen moves freely over the drawing surface.

One could, of course, argue that the drawing process is never completely unconscious. Conscious control, however, can often be lost in the intensity of the repetitive gesture, as the drawings appear to take on their own shape or 'life', so to speak, or to possess characteristics that are different from what might be achieved by a more premeditated or planned act. This I see as an immersive experience of 'losing myself' in the process. The looping of the action essentially acts to narrow the perceptual focus, and by doing so it ensures entry into a state of dissociation or trance. The densely layered ambiguous markings that manifest as the final results of the drawing process, whilst not the ultimate goal of that process, are the intriguing physical traces of the medium. These traces, in their intricacy, density, and convolution, can suggest all kinds of phantasmatic forms to the open, pattern-forming mind.

4.5 *Drawing Phantoms* Performance

As suggested above, this personal and regularly enacted drawing ritual served as the basis to develop a solo live work, which morphed into the *Drawing Phantoms* performance over time. *Drawing Phantoms* is a sonic/physical performance that involves the tracing of shapes onto markable surfaces (e.g., paper), having this action amplified by piezo-disks or contact microphones, exploiting all the media involved to manipulate the perceptual auditory image and to 'lure out' sonic phantoms. As the performance is essentially built upon repetition as a device to construct and deconstruct meaning, it requires an extended duration (in the range of at least thirty minutes) for successful power and effect. Over that span of time, the effect produced is one in which our natural inclination towards pattern recognition produces a special class of sonic phantoms that resemble human voices (we also admit the possibility of

---

[37] This term coined by William Carpenter, a physiologist in the mid-1800s, was used to describe how, when we expect a movement and we have an idea of it in mind, this expectancy can actually cause our muscles to unconsciously produce the expected movement. This concept was used to explain, for example, how the Ouija board works and also how automated writing feels like the same external force is driving the movement of the pen. See Ondobaka and Bekkering, "Hierarchy of Idea-Guided Action and Perception-Guided Movement."

listeners hearing voices that they attribute to other more ambiguous entities). Despite the relentless repetition involved in the processes that comprise *Drawing Phantoms*, for me the fascination comes from the emergent patterns that appear in the foreground, disappearing randomly and spontaneously, providing endlessly changing listening perspectives (this is true for me both as a practitioner and as a listener).

The drawing sounds are produced by the rhythmic repetition of continuous movement, using graphite and paper on the amplified surface. The circular and curvilinear figures are driven by the movement of the arms and hands while drawing, and so sonic, visual, and somatic properties are of equal importance in determining the personal outcomes of the performance. In the drawing of the circular and curvilinear figures, a significant expenditure of physical energy is necessary to maintain the rhythm, along with an equally intense demand upon the cognitive apparatus to preserve the continuity of what eventually becomes a semi-automated process. The performance generates several looping auditory patterns, corresponding to the changing visual shapes.

In terms of the specific materials used for the *Drawing Phantoms* performance, I worked with sufficiently large sheets of paper (at least A1 or A0 size) cut from recycled and robust paper rolls, such as large wallpaper rolls. I have experimented with all kinds of marking tools, including pencils, charcoal pens, wax and oil crayons, markers, 'invisible' markers, and paint. In performance, I find it effective to use fine-tipped graphite pencils (due to their high durability and hard-edged nature) for the first part of the performance, which usually consists of marking circles in a unilateral one-hand drawing mode. I then switch to wax crayons for the second part of the performance, which is a bilateral set, two-handed drawing with multiple simultaneous transition shapes.

Although I usually begin with movements that would result in a circular shape, I have also experimented with other trajectories, transitioning between various shapes to generate different rhythmic and timbral structures. Drawing circles and curvilinear shapes in repetition, though, is a simple and effective way to create strong visual and auditory patterns without requiring too much premeditation upon the results of the act (such as anxiety over whether or not one will be 'successful' in rendering a complex figure that is an adequate enough physical manifestation of a mental image). Despite the fact that the curvilinear patterns have a stable, rhythmic structure, there will always be timbral variations that owe themselves to the multiplicity of factors involved with the drawing process: the thickness of the line, the speed of the drawing, or the angle and pressure of the contact point of pencil and surface. The latter can easily be altered so as to affect the accentuation of the emerging sound patterns. Changing the grip of the graphite has a perceptible impact on the sound of the patterns, and this can be used as a technique to modify the timbre of the sounds and to bring out or accentuate or illuminate the phantom 'voices.' These voices can then further be highlighted and brought to life with equalization techniques applied to the amplified sound. This specific accentuation of the patterns with the aim of producing sounds that are convincingly voice-like is somehow equivalent—albeit by very different

Figure 4.5 Live performances with real-time video projection of versions of the compositional series *Drawing Phantoms*, presented as *Drawing EVP* at Störung #9 Festival, Barcelona, Spain, 2014 (photograph by M. A. Ruiz © 2014 Störung) and *Drawing OVP* at Volumens Festival Internacional de Exploración AudioVisual, Valencia, Spain, 2015 (photograph © 2015 Volumens).

means—to the tradition of 'Electronic Voice Phenomena' (a comparison that will be elaborated on in a later section).

It should be reiterated here that the physical practice requires intense focus and energy to produce strong, driving drawings, and to slowly engage the entire body in the movement. Therefore, some additional amplification of the physical effects of this practice would be useful. Not all of them are, interestingly, associated with physical fatigue: after a typical duration of five minutes or so, the hand and arm movements feel powerful and automated, and then the body seems to instinctively adapt to the rhythm of the movements. Fluid motion and symmetry are vital for the development of the rhythmic structure and a continuous flowing movement. I tend to work with strong lines with even, regular patterns whilst simultaneously playing with subtle variations. Playing with the range of speeds might also produce markedly different effects, and so attention needs to be paid to one's sense of timing. I have played with multiple layers of variations and transformations of circular and non-circular patterns in addition to simplifying the process in order to focus on one at a time, extracting one shape and working with it tirelessly to generate the desired amount of sonic phantom patterns.

Bilateral action (in this case, working with two hands with different movements), also serves to intensify the sonic output and strengthen the pattern-making process visually, aurally, and physically. As might be anticipated, this type of movement can literally double the intensity and complexity of the patterns and increase the forceful impact of the performed piece dramatically. Depending on the overall proposed duration time of the performance, I plan and 'compose' the trajectories and transformation patterns of the drawn shapes in different sequences over time. Each performance, therefore, will change in these specific terms as well, depending on the circumstances. The entire process in performance typically leads to the creation of one continuously interwoven curvilinear line drawing. At the end of the performance, the page, initially blank, is then covered with densely drawn layers of intricate curvilinear patterns.

4.6 Expanding by Narrowing—Everyday and Induced Trance

In a keynote paper delivered to the American Federation of the Arts' annual conference in 1957, art historian Meyer Schapiro[38] issued a stern warning against the otherwise innocuously named 'arts of communication.'[39] This came at a time when 'communication' had been reconfigured as a buzzword for the nascent field of cybernetics, bringing with it Rudolf Arnheim's assumption that the work of art could apply to advancements in techno-science as well as in those areas more traditionally recognized as art. Schapiro's complaint was that these arts of communication, a fusion

---

[38] Meyer Schapiro (1904–1996) was significant to the development of art history and criticism in America, also having some influence on American modern artists such as Pollock and de Kooning.
[39] Schapiro, "The Liberating Quality of the Avant-Garde."

of art and technology, would essentially represent a triumph of mass media over the more irrational spaces of life. In such a condition, "the greatest amount of information would be delivered to the largest possible audience as efficiently as possible," with the ineluctable result being that "content was aimed at the lowest common denominator in order to reach as many people as possible; the media embodied the debasement of art by cybernetics."[40] Schapiro was not an anomaly then, and certainly is not now, in casting a wary eye upon maximalization of communication and the effects that it would have upon creative endeavors. Schapiro's grounding in art history was also strong enough that he must have known, when making these statements, such a project would fly in the face of a practice of willful information *reduction* that had informed the evolution of creativity for millennia.

This form of meaningful—even necessary—reduction for personal significant transformation appears in all its guises time and again. Take, for example, Dada-aligned artist and philosopher Hugo Ball,[41] who, in the diaries of his most turbulent years, once declared "we have to lose ourselves if we want to find ourselves."[42] While Ball's declaration was aimed more squarely at those participating in intellectual pursuits, he surely must have been aware—by way of his dedicated studies in mystical and ecstatic practices—that its content could just as easily speak to a deeper tradition of deliberately reducing the number and variety of concurrent sensory impressions that we receive in order to broaden and deepen our total knowledge. Most, if not all, key anthropological texts on shamanic practice acknowledge the importance of techniques that will successfully combine a "focusing of attention in one area ... accompanied by a reduced awareness of surroundings outside this focus."[43]

This is a concept that has not been extinguished even in an age that seems, superficially, to downplay the value of archaic methods of self-inquiry. The influence of the trance state upon human creativity is a story that spans millennia, and covers perhaps every region of the globe in which human societies have existed. As such, its persistence to the present day should come as no surprise, despite our lives now being qualitatively different from those of hunter-gatherer societies in which shamanic influence once held sway. Many creative undertakings that seem to be conducted within a thoroughly modernized environment still find themselves evoking the potency of the trance state as a developmental tool—as was the case with the neo-shamanic approach of the late post-industrial percussionist Z'ev—and the same can be said of artistic activities operating more explicitly under the aegis of techno-science. This latter set is exemplified by a recent musical performance piece based on the audio realization of brainwave activity as registered by participants wearing EEG headsets (itself not too different in spirit from Alvin Lucier's 1965 *Music for Solo Performer*,

---

[40] Kaizen, "Steps to an Ecology of Communication."
[41] Hugo Ball (1886–1927) was one of the principal players in, and chroniclers of, the Zurich Dada art movement, while also being one of the prototypical 'sound poets.'
[42] Ball and Elderfield, *Flight out of Time*.
[43] Vitebsky, *Secrets of the Shaman*.

the first widely documented piece to 'musicalize' such equipment). The press copy accompanying the recorded souvenir explains that "the performer develops and gains control over his own brainwave patterns during the performance, by listening to the sonic results of his mental activity...by changing electric impulses of the brain *during meditative and ecstatic trance* [italics ours], the performer begins to sculpt the sound through the mind alone."[44]

Another prominent example of the modern persistence of 'expansion through reduction' comes to us via the oeuvre of the video artist Bill Viola.[45] In the process of making a comparison between the way in which his chosen medium operated vis-à-vis commercial television, Viola proposed that video allowed creators to show "less and less information in more and more time," and in doing so managed to flip the ethos of commercial advertising and the hyper-speed programming of information-saturated networks like MTV on their heads. Viola's proposition regarding the paradoxical increase in meaning via reduction of broadcast information is intriguing, not because it is a recent development in the history of perception, but because it rehabilitates a sensibility that had served humans in antiquity: what Viola proposes is, after all, the guiding principle behind the trance state (his conjuring of this via video is also an important reminder that trance is achievable irrespective of medium). "One can go into trance without music; and can listen to music and not go into trance,"[46] and both conditions may be determined culturally as much as they are psychoacoustically.

Of course, the conjuring of such states by shamans may have called for significantly more physical and mental exertion than the process of patiently watching a long-form video, particularly given the much higher stakes involved: see, for example, Vitebsky's insistence that shamanic power itself "depends on keeping control of the trance state,"[47] while fantastic results such as soul travel to either the heavens or the underworld implied a result in keeping with rigorous, self-imposed asceticism or extreme exertion. However, the essential concepts of 'losing ourselves to find ourselves' and of 'expanding through narrowing' apply in all cases.

Before proceeding further, it may also be useful to clarify key misconceptions about trance, as this has become a hopelessly loaded term. As Becker cogently warns, trance, "like most natural language categories, is a cover term for a set of things that more or less resemble each other."[48] This is borne out by the fact that trance has a pejorative meaning in modernity, which may have been inherited from Victorian physicians' concept of trance as a medical condition that we now know more properly as cataplexy. The modern version of this diagnosis has more to do with describing a state of helpless passivity or subjugation to societal forces with the more positive quest for a renewed and heightened sense of awareness and selfhood (consider how expressions such as

---

[44] https://fragmentfactory.bandcamp.com/album/wellenfeld. Accessed October 25, 2019.
[45] Bill Viola (1951–) is a video artist whose lyrical experimentation with slow-motion sequences is one of the more notable features of his body of work.
[46] Becker, "Music and Trance."
[47] Vitebsky, *Secrets of the Shaman*.
[48] Becker, "Music and Trance."

'consumer trance' retain as much currency as other esoterically tinged pejoratives like 'zombie' when one is attempting to condemn a willful lack of agency).

Conversely, anyone who has witnessed ritualistic trance states first-hand will likely attest to the fact that they are the antithesis of this sort of isolated passivity and stasis, involving as they do frenzied dancing, noise, and public communion (and this is even true of those states that may occur with no apparent stimulus, particularly the so-called 'savage' trance of Nigerien Bòorii adherents that comes about "unexpectedly, without any music having been played").[49] So, it would be wise for the purposes of this discussion to jettison any moral preconceptions based upon the circumstances surrounding the trance, and to instead follow Becker's realization that "one of the common characteristics of the trance category is its focus; its intensity" as well as the secondary realization that "trance is not a digital on-off state... there can be many degrees of trance."[50] Taken together, a picture unfolds in which we can see common features in trance states that occur in clearly 'ritualistic' contexts and those that arise from more quotidian means, up to and including the most readily accessible communications technologies.

In my own work, one aspect of particular relevance is the intent to create a powerful and transformative experience for both the listener and myself. Becoming aware of the techniques for entering into and deepening trance-like states has given me a significant insight into a process that was, historically speaking, intuitive and mostly unconscious. Creating such states allow for the introduction of a feeling of heightened emotional experience, and, in keeping with Ball's pronouncement above, enable us to 'lose ourselves' in the creative process (in this context, it is useful to know the roots of the word ecstasy in the Greek *ek stasis*, 'standing outside oneself,' when understanding why this state is often the promised and desired afterglow of trance activity). This kind of feeling is one that is extremely difficult to provide a verbal record of, and indeed Becker outlines the challenge of doing so when detailing how the basis for such is at the neurological level rather than the linguistic (to wit: "ineffability, at the neuronal level, would seem to relate in part to the inhibition or shutting down of sections of the language areas of the brain and some of the connections between the cortex and the hippocampus... that establish short-term memory").[51] Nonetheless, and in spite of the myriad ways in which the trance state is likened to a disembodied flight or a reduction to pure vision,[52] the experience of such is very tangible.

In *Drawing Phantoms*, the audibly repetitive motor activities operate as techniques of trance induction. In this case, the curvilinear shape is the basic fundamental primary pattern. It becomes a way to intensify the cognizance of internal processes and to narrow down what might typically be described as a 'floodlight' field of perception

---

[49] Erlmann, "Trance and Music in the Hausa Bòorii Spirit Possession Cult in Niger."
[50] Ibid.
[51] Ibid.
[52] The latter is described poetically by an Inuit shaman in Mircea Eliade's landmark work on the subject: "My body is all eyes/Look at it! Do not be afraid!/I look in all directions!" See Eliade, *Shamanism*.

to a more acute 'spotlight' focus on the intensified sounds, whilst simultaneously excluding external or peripheral factors. Limiting choice with minimal elements establishes an important condition for trance induction. The repeated loop pattern, with its phantom-inducing potential, is a necessary condition for this trance-inducing process. With dissociation brought on from heightened states of absorption comes the potential for phantom perception and illusion generation.[53] These are temporary liminal zones where we may experience strong perceptual alterations or illusions, such as hearing 'voices' or struggling to distinguish between different versions of apparent reality.

Today, when most people think of trance, they associate it with some kind of exceptional, hypnotic state of mind (and even the derisive 'consumer trance' noted above suggests an exceptional degree of involvement in unexceptional activity). According to Dennis R. Wier, director of the Trance Research Foundation, the state of trance that we normally associate with unusual or altered special situations is more common than we think, being in fact "as common as grains of sand."[54] Simply stated, quotidian life is every bit as capable of this level of sensory involvement as are the loftier, more romanticized forms of trance induction via devotional or spiritual activity identified as 'institutionalized forms' of trance by Richard J. Castillo.[55] According to Wier's vision, under such common circumstances as daydreaming, listening to music, surfing the internet, trying to solve a conceptual problem or watching television, there is a good chance that we are effectively in a trance.[56] In general terms, his trance theory provides a useful framework for clarifying what are perhaps the essential conditions and techniques required for all trance-inducing processes. More specifically, it has been useful for me to shed light on my own working process with my use of changing looping structures within the framework of ritual and repetition as techniques to induce the phantasmatic in my music.

A trance state, then, can be subtle, or deeply intense and more extreme, depending on the conditions, and particularly when the focus is placed upon a singular task or percept to the apparent exclusion of all other peripheral information. Most specific techniques that aim to induce trance states (for example, meditation or hypnosis) work on this very basis of narrowing the attentional field by focusing attention on a single sound, an image, a movement, and so on.[57] I consider the state of trance to be on the extreme end of a spectrum of music listening, of which the other end is what we would call a 'normal' listening experience. Under this vision, trance is fundamentally not a different phenomenon than mundane listening, but instead an intensification and amplification of it. In the state of trance, our perceptual processes seem to operate at a plateau upon which sound is lived and inhabited and not just simply 'heard.'

[53] Hilgard, "Divided Consciousness and Dissociation"; Brown and Volgsten, *Music and Manipulation*; Pec, Petr, and Raboch, "Dissociation in Schizophrenia and Borderline Personality Disorder."
[54] Wier, *Trance*, 24.
[55] Castillo, "Culture, Trance, and the Mind-Brain."
[56] Wier, *Trance*; Wier, *The Way of the Trance*.
[57] Baker, *They Call It Hypnosis*. See also Aldridge and Fachner, *Music and Altered States*.

Lewis-Williams outlines how trance states may be induced by a variety of means; for example, using sound as a driving force: the phenomenon he describes as 'auditory driving,' closely related to that of entrainment,[58] is understood as the situation in which there is a feeding of the auditory cortex of the brain with a sound stimulus, in such a way that the neural activity in the auditory cortex synchronizes and resonates with the input. Auditory driving (also known as 'sonic entrainment') has since become a popular concept among students of trance states, given its implication that music has direct physiological effects on the brain both during and before trance states occur. This effect can be produced by means of persistent repetitive sound loops, like those that take place in drumming, and in fact drumming remains the prime instrument identified with this activity, given the enduring and sometimes exclusive association between drums and trance-like behavior.

Reports of the fantastic, superhuman behavior following from drum-led rituals, for example, ones in which "participants in drum ceremonies [endure] ritual ordeals that would ordinarily be extremely painful,"[59] continue to captivate the imagination, particularly as scientific data lend credence to the fact that rhythmic stimuli occurring in tandem with pain can have an inhibitory effect on the latter. Likewise, musicologist Kathryn Vaughn also describes an example of entrainment in Karelian singing, wherein the synchronization of vocal cords' vibrations to that of more visible bodily gestures provide the singer with a sensation of the spirit entering the realm of the dead.[60] Keeping these examples in mind, it is also easier to understand the unflagging popularity of marathon electronic dance music events, in both their mainstream and underground variants. This also testifies in part to an ongoing belief in the transcendent power of intensely percussion-driven rhythmic music: being that percussion is, per Walter and Walter, "a sound source stimulus containing components of supraliminal intensity over the whole gamut of audible frequencies,"[61] it has a clear ability to maximize the stimulation of the whole basilar membrane and, it follows, to stimulate greater, more frequent and more intense emotional responses.

Though it has already been implied above, it is important to note that sound sources other than percussive ones are capable of achieving this effect. As Lewis-Williams attests:

> Audio driving, such as prolonged drumming, visual stimulations, such as continually flashing lights and sustained rhythmic dancing, such as among Dervishes, have a similar effect on the nervous system. We also need mention fatigue, pain, fasting and of course, the ingestion of psychotropic substances as means of shifting consciousness along the intensified trajectory towards the release of inwardly generated imagery.[62]

---

[58] Lewis-Williams, *The Mind in the Cave*.
[59] Neher, "A Physiological Explanation of Unusual Behaviour in Ceremonies Involving Drums."
[60] Vaughn, "Exploring Emotion in Sub-Structural Aspects of Karelian Lament."
[61] Walter and Walter, "The Central Effects of Rhythmic Sensory Stimulation."
[62] Lewis-Williams, *The Mind in the Cave*, 54.

At least this much is also suggested by Gilbert Rouget in his own criticisms of Neher's auditory driving theory, particularly in Rouget's contention that "practically any instrument may be used... from a nearly inaudible zither to a powerful battery of drums." Meanwhile, similar driving effects in other sensory modalities—such as the kinesthetic or tactile—can also be achieved. This is the case with repetitive flashing lights and so-called 'photic driving,' where continuous rhythmic movement causes an intense focusing on a single point and has a powerful effect on the nervous system. Interestingly, photic driving has applications that extend beyond the purely artistic, as it has been successfully utilized to aid in treatment of depression and to diagnose epilepsy and tumors.

Having noted this, a brief diversion into other artistic media may provide some additional valuable amplification of key facts that directly relate to the effects and experiences described for the *Drawing Phantoms*. Just as a plethora of musical instruments can provide the above effects, it should also not go unsaid that rhythmic stimulation across multiple sensory modes greatly heightens the potential for a combination of deep emotional responses and involuntary physical responses. It explains the efficacy of audiovisual works such as Tony Conrad's groundbreaking structural film *The Flicker*, or phenomenologically similar films from Peter Kubelka[63] (*Arnulf Rainer*) and Paul Sharits (*Ray Gun Virus, N:O:T:H:I:N:G*). When examining Conrad's stroboscopic film in particular, we can see how the same alpha rhythm stimulation that Neher[64] cites as the key to 'auditory driving' also contributes to other profound transformations in perception and proprioception. One such transformation, which Jack Burnham[65] saw as heralding a transition from an 'object-oriented' culture to a 'systems-oriented' one, is sufficiently similar to the phantasmatic effects encountered through audial means of entrainment. Namely, we begin to perceive a change that "emanates not from *things*, but from the way *things are done* [emphasis in the original]."[66] Sharits himself elegantly echoed this statement when saying "flashes of projected light initiate neural transmission as much as they are the analogs of such systems... the human retina is as much a 'movie screen' as the screen proper."[67] When this technique is transferred back into the audio realm, it presents great potential for creating works that, though they may initially be *about* a certain type of transformation, graduate from that state of representation to simply *being* transformation.

Again, this seems to bring us back to the *ekstasis* of the trance state; the ability to step outside of customary patterns of behavior long enough to discern internal and environmental workings whose presence had been obscured by a daily accretion

---

[63] Peter Kubelka (1934–) is an avant-garde filmmaker and theorist, as well as curator of the Austrian Film Museum.

[64] Neher, "Auditory Driving Observed with Scalp Electrodes in Normal Subjects"; Neher, "A Physiological Explanation of Unusual Behaviour in Ceremonies Involving Drums."

[65] Jack Burnham (1931–2019) was an American writer on art and technology, who taught art history at Northwestern University and the University of Maryland. A main force behind the emergence of systems art in the 1960s.

[66] Burnham, "Systems Esthetics."

[67] Sharits, "General Statement for 4th International Experimental Film Festival, Knokke-Le Zoute," 13.

of superficialities and minutiae. As it is with apprehension of sonic phantoms, this process perhaps clues us in to the fact that, the more familiar we become with habitually received sensory information, the deeper that knowledge sinks into consciousness and the less we are aware of even possessing the knowledge that we do possess. Simply being aware of those different depths or multiple layers of consciousness can be a powerful means towards expanding the possibilities for communication and creation. Put another way, when critic Rosalind Krauss states of the flicker film phenomenon that it "was invented to stop time,"[68] this, too, implies a kind of positive stasis in which distractions and inessential sensory phenomena are made to fall away long enough to provide a fascinating cutaway view of the machinery of reality. In addition to such potential benefits, exercises in 'driving'—auditory or otherwise—can bring us to a self-reflexive state, one in which sonic art becomes effectively 'communication about communication,' projecting information about its making and our perception of it at the same time that it is being projected into the world.

4.7 Loop Multiplicity

The state of trance dissociation experienced during the performance of *Drawing Phantoms*—as it would be the case of other similar intensely repetitive performances—is of course subjected to a number of potential distractive interferences. It can be easily disrupted, with the sudden termination of all the illusions that participants worked to attain. In order to strengthen and maintain the experience, I have experimented with the use of multiple loops, which can be used during a performance for added complexity and power, maintaining the stability of the flow, while also deepening the state of absorption.

I have experimented with several expanded versions of *Drawing Phantoms*, combining the amplified drawings with constructed mechanical-sounding loops of instrumental sounds in live performance to create an immersive sound environment with multiple trance loops. Stemming from the work I had done exploring the instrumental looping patterns (particularly in the *Harp Phantoms* compositional series), I digitally created layers of shifting, looping patterns to give rise to emergent and ambiguous sonic results, which I then combined live simultaneously with the *Drawing Phantoms* ritualized performance. The instrumental patterns were composed by layering plucked string recordings to produce precise results evocative of mechanical processes. The next step was to duplicate these patterns and, with some variations and transformations, layer them together in interlocking fashion, to create constantly alternating 'figure-and-ground' focus points. This was based on the idea that it could be interesting and effective to combine the relentless and driving force of the instrumental looping patterns as a way to entrain both sounding sources in the live drawing process. The phasing and colliding patterns in combination with the drawing patterns create what I have before described as 'in-between moments.'

---

[68] Krauss, "Paul Sharits."

This term, in my personal nomenclature, refers to those threshold-crossing moments where the sonic texture is at its most ambiguous from a perceptual perspective. Transition phases when patterns are slowing and moving at different speeds against each other create fruitful ambiguity. During these liminal moments, there is potential for listeners to attend to constantly shifting new patterns emerging to the foreground. In the situation when patterns are identical and phased slowly against each other we quickly hear a variety of different sub-streams and sonic phantoms, due to the auditory stream segregation process. As patterns develop and continue evolving, one shifts awareness between these multiple changing states. Each state has its own trance-inducing potential and it is the listening to the combination of states that keeps one engaged. The potential of these in-between ambiguous spaces to spontaneously facilitate the generation of various patterns beyond the control of the listener is one of the fundamental elements I find particularly interesting in this process.

4.8 *Drawing Phantoms* Séance

As described above, my own performances of *Drawing Phantoms* are carried out with a strong ritualistic component. In a further evolution of this series of pieces, I have worked in collaboration with Dutch performer Nathalie Smoor,[69] incorporating more elaborated theatrical and movement elements, to propel this performance ritual into the domain and the tradition of the spiritualist séance.

In this version of *Drawing Phantoms*, the stage scenery of the ritual is carefully set up in its details with the intention of intensifying the performer focus and immersion on the sound and the visual results that she is producing as she performs. All of this is, of course, strongly modulated—in the pace of actions, in the stage presence, in the lighting, even in the performer's clothes and hairstyle—by the séance nature of the performance. The performer wears a white Victorian-style dress, her long dark hair down in marked contrast over the dress. She stands in front of the white blank page attached to the table and the scene is illuminated from above with black ultraviolet light, increasing the contrast between white dress and black hair, showing only her figure and the table, and darkening everything else around. Over the course of the performance (with an approximate duration of thirty to forty minutes), an additional zenithal spotlight gradually increases in intensity, seamlessly blending with the black light, and reinforcing the visual presence of the drawing that is progressively taking shape from the actions of the performer.

This light and graphic progression is naturally accompanied with a sonic crescendo of increasing complexity, intricacy, and inducing power for the generation of sonic phantoms: the 'voices' from the séance. In this version of the performance, I have also explored transforming the incoming drawing sounds in real time by means of various digital signal-processing techniques. These transformations are layered upon the live amplified drawing sounds and the overall output mixed and spatialized over a

---

[69] Nathalie Smoor is known for her work in theatre, mime, film, music, dance, and martial arts.

Figure 4.6 *Drawing Phantoms* séance. Live performance by Nathalie Smoor at Handmade Homegrown Festival, Dakota Theatre, The Hague, The Netherlands. Photographs by Barbara Ellison, 2013.

quadraphonic sound system set-up around the performer and the audience. Multiple and diverse 'voices' are indeed invariably 'heard' throughout this séance.

From my experience in presenting this piece internationally, it is quite common after such a performance that audience members with different native languages report hearing all kinds of phrases in their respective languages. Listeners are often convinced that there are 'secret' recordings of real voices embedded into the performance sound. The process of hearing the sonic phantom voices made manifest by the drawings is, of course, the product of our apophenic brains; or their acting in a capacity as 'prediction engines' that attempt to match new, incoming information with those aspects of worldly experience that we already know to be true. In addition, audience members are completely primed to hear 'voices' from the very start (owing to the title and description of the piece), while the mysterious atmosphere created in the space with this particular 'ghostly' Victorian setting naturally contributes to that effect.

In the context of apophenia, the recognition of these 'voices' is a typical cognitive misinterpretation of—or, perhaps more precisely, reaction to—random broadband noise (the general category that includes so-called 'white,' 'pink,' and 'brown' noise, but also other irregular versions of a dense frequency spectrum audio signal). The sonic result of the drawing actions has such a broadband nature, with a mass of multiple simultaneous audible frequencies.

The effect is particularly powerful, for obvious reasons, when we are primed to expect that we will hear mysterious voices or any other implicit or indirect suggestion of that kind. It is nonetheless surprising to see how easy it is for our brains to come to interpret such noise patterns as words once we are provided with the suggestion that we might indeed hear words coming out of that apparently meaningless noise.[70] James E. Alcock provides some elaboration upon the cognitive processes that undergird all of this, describing how our brains are guided to a great degree by what we expect to hear:

> If you are trying to hear your friend while conversing in a noisy room, your brain automatically takes snippets of sound and compares them against possible corresponding words, and guided by context, we can often "hear" more clearly than the sound patterns reaching our ears could account for. Indeed, it is relatively easy to demonstrate in a psychology laboratory that people can readily come to hear "clearly" even very muffled voices, so long as they have a printed version in front of them that tells them what words are being spoken. The brain puts together the visual cue and the auditory input, and we actually "hear" what we are informed is being said, even though without that information, we could discern nothing. Going one step further, and we can demonstrate that people can clearly "hear" voices and words not just in the context of muddled voices, but in a pattern of white noise, a pattern in which there are no voices or words at all.[71]

---

[70] Sacks, *Musicophilia*.
[71] Alcock, "Electronic Voice Phenomena."

As regards this, the example provided by Merckelbach and van de Ven[72] is especially amusing, bearing similarities to a sort of absurdist comic scenario: in the process of auditioning a recording for students that contained nothing but white noise, the experimenters suggested that the sounds of Bing Crosby's *White Christmas* were lurking somewhere beneath the auditory threshold. A full 32 percent of respondents 'confirmed' that the holiday evergreen was indeed present, despite the fact there was no trace of Crosby singing *White Christmas* at all in the recording, just white noise. This is just one among a host of examples that showcase the truth of apophenia; that our brains are guided to an overwhelming degree by what we expect to hear and in that example expectations were guiding participants to hear a signal in the noise that was not there.

Though we will touch upon a more colorful and psycho-socially intriguing manifestation of this in a moment, for now let us turn our attention to the bistable phenomenon known as 'sine-wave speech,' in which intelligible phonetic information arises from a combination of multiple time-varying sine tones. This provides an excellent representation of the predictive mechanisms and their active generation of communicative content (or, at the very least, of content that is aesthetically intriguing enough to affirm some aspect or another of the perceiver's existence). This subspecies of sonic phantom is particularly worth contrasting with more noise-based forms, since the sine wave represents the simplest and least chaotic form of periodic acoustic excitation, at the opposite end of the complexity scale from noise. Yet the mechanisms for 'phantom' production are largely the same, with sine-wave speech being a sonic experience that does not in any way resemble the acoustic properties of speech (the examples typically provided often sound like 'keening whistles' with different pitch and volume properties). Yet, even with the spectral qualities of the sinusoid combinations being vastly different from those associated with speech, with nothing comparable to the articulations provided by the human mouth (i.e., its aspiration and frication), these skeletal examples can be interpreted as vocalizations once listeners are prodded to listen to this *as if* they were a synthetic form of speech. Once listeners have assented to there being speech, the door is wide open for apophenic interpretations of just what the precise content of that speech may be. Furthermore, if listeners hear the actual original sound file (from which the sine-wave version is created) and then they re-listen to the sine-wave version, the sine-wave file now sounds completely intelligible and the precise sentence being spoken is clearly audible. Andy Clark explains that this is due to listeners now having access to the correct top-down model, or the right expectations, which are strong enough to cause the hearing of signals that are not there in the original transmission:

> Its like hearing a familiar song when it's played in the shower or on a bad radio receiver... It's a very striking effect and experience... It gives you are a real sense of what is happening when a predictive brain gets to grips with the flow of sensory information.[73]

---

[72] Merckelbach and van de Ven, "Another White Christmas."
[73] Clark, "Perception As Controlled Hallucination." See also Clark, "Whatever Next?"

I have deliberately provided such apophenic suggestions in some program text descriptions of *Drawing Phantoms*, or even in variations of the title used for different performances, such as *The Other Voice* or *Drawing EVP*. Under such implicit exhortations, we then assume we will hear voices, and this establishes an expectation that in turn makes us hyper-alert to detect such vocal patterns. Our brains are 'keyed up,' so to speak, to pattern-match the noisy sounds of the drawings with words that we are expecting to hear. The relentless repetition of the drawing loop will force us to intensely 'tune in,' allowing the random noisy sounds to monopolize attention. This provides the time necessary to give the pattern-forming brain the chance to 'create' voices that are not actually acoustically present in the incoming signal. It is a dynamic process of auditory apophenia in action. Moreover, our gradual immersion into a temporally extended performance environment intensifies our 'voice-expectant' mental state, temporarily and partially disabling our critical faculties once dissociation takes hold. We therefore hear such semantic structures where none exist. Once we cognitively 'hit target' and start hearing vocal patterns, further exposure to sounds will reinforce these sonic phantoms in our perceptual system. Among the vocalizations that arise, some words are uniquely reported by representatives of certain cultures, and, as mentioned before, particularly in their host languages: a fact that lends additional credence to the relevance of cultural framing to apophenia.

4.9 'EVP' and 'OVP'

The theory of our perceptual systems' 'global optimized performance'[74] suggests that illusions, sonic or otherwise, still represent the most accurate possible 'real world' sources for those percepts. Nevertheless, our desire for attributing ambiguous sensory materials to non-real world sources is considerable, and when we are primed or invited to do precisely that, there are few limits to the variety of phantasmatic gems to be mined. The previously discussed top-down model of information processing is not always foolproof, and on those occasions where it does fail, our personal beliefs can cause us to see any information matching our hypotheses about a given sensory picture to be true. This form of perceptual play becomes all the more prevalent where emotional stakes are high, up to and including situations in which we have been primed to— for example—hear the voices of long-lost loved ones or to hear the advice of sidereal entities which had, up until the point in which we perceived them, completely avoided detection by our most advanced technologies.

Nowhere are such tendencies, as well as the power of suggestion for auditory apophenia, as clear, vivid, and consequential as in the realm of so-called 'Electronic Voice Phenomena' (EVP). EVP, whose history begins in the 1920s, is allegedly recorded communication with ghosts, spirits, or other paranormal entities, by means of audio recorders and various other electronic devices, such as radio receivers.

[74] Lupyan, "Cognitive Penetrability of Perception in the Age of Prediction."

The modern work with EVP derives largely from the combined efforts of erstwhile Estonian documentary-film producer Friedrich Jürgenson (1903–1987) and Latvian parapsychologist and Jungian psychiatrist Konstantin Raudive. Their respective texts *Sprechfunk mit Vestorbenen* ('Voice Transmissions with the Deceased') and *Unhörbares wird Hörbar* ('The Inaudible Becomes Audible,' later published in English as *Breakthrough*) have provided the theoretical and practical mooring for later experiments in this field (to say nothing of Raudive's archive allegedly consisting of thousands of EVP recordings).

Jürgenson had his personal epiphany when playing back birdsong tape recordings made earlier, and hearing a man's voice commenting upon the birds' nocturnal singing patterns (which would soon enough come to the attention of Raudive, with the two actually working together for a period directly after the publication of Jürgenson's first book). Jürgenson was completely intrigued by this outcome, which led him to dedicate much time to making recordings in quiet places, devoid of humans and other noises, places where there was 'nothing.' Jürgenson kept detecting voices in his recordings, and so he undertook what he considered to be a systematic scientific long-term study of this phenomenon, gathering and analyzing many recordings; the results of this research were presented in his 1964 book *Rösterna från Rymden* ('Voices from Space').

Others who are not as widely recognized as Raudive and Jürgenson nevertheless have charted a parallel course: the American Attila von Szalay published notes on his findings in *The American Journal of the Society for Psychic Research* in January 1959, mere months before Jürgenson's revelatory bird commentary recordings (and some might credit Szalay with being the true trailblazer here, since he had been active since the early 1940s, recording first on a record-cutter and later on a wire recorder before he began experimenting with recording onto magnetic tape). Today, EVP has become a realm of specialized interest with perhaps thousands of practitioners worldwide and seemingly innumerable websites devoted to the practice.

Raudive's method, as explained in *Breakthrough*, involved means no more sophisticated than a tape recorder placed in a room near a white-noise generator (the most accessible of which would be a radio tuned to a non-broadcasting frequency). Upon doing this, the interlocutor would ask, in the same manner as in a spiritualist séance, if any 'unseen friends' or acquaintances were present, after which any number of follow-up questions could be posed. Interest in this technique has become widespread enough, since the 1960s, to support numerous international organizations dedicated to ongoing research. Nowadays the EVP enthusiast tends to set up equipment within an emotionally resonant location likely to bear some 'imprint' of past trauma, or more generally popular classic 'spooky' locations such as graveyards or haunted houses, and records whatever is found there (often by leaving the recording equipment unattended overnight). A less 'site-specific' variant of the EVP practice extracts these recordings from endless sessions of scanning the radio waves with a receiver. Later on, these recordings are played back, listened to carefully (and while in an apophenic frame of mind), and analyzed to detect voices or other communication purported to be from the world of the supernatural. The recordings one typically encounters upon playback are highly noisy, which seems the inevitable result of boosting the sensitivity level of any recording microphones to maximum. Even in an environment where there is supposedly nothing to record, this signal boosting will unavoidably leads to a noisy result, and hence EVP. The heightened levels of noise, in turn, require systematic playback repetition of specific sections of the recordings. Altogether, these steps comprise perhaps the most widely deployed technique for finding the fragmentary EVP voices (incidentally, the incomplete nature of the received 'communications' often goes unexplained, and their initial indiscernibility is occasionally chalked up to factors such as the entities' supposed penchant for polyglot speech—more on this in a moment).

EVP practitioners claim to have thousands and thousands of recordings of ghosts, spirits, and other supernatural or extra-terrestrial entities. Notwithstanding the fact that some of these 'voices' are in fact voice transmissions from currently existing, fully corporeal entities (i.e., from interference bits of radio broadcasts and mobile devices picked up by the receivers), they all can be readily explained by auditory apophenia. Despite the complete lack of any real evidence on anything paranormal in these recordings, it does not take much for enthusiastic minds to hear voices in a multitude of repeated fragments of recordings. Professor James E. Alcock describes the self-reassuring nature of the prototypical arguments of the EVP hunter:

> Some voices of spirits or entities are very close to the background level of static; others may be clearly heard. If the speech is difficult to understand, remember that the spirit talking may be talking in a language or dialog that is not in common usage today. The voice can also be in reverse, you would need a computer to reverse this to hear it.[75]

The mention of a computer here suggests an evolution of different EVP iterations that proceed apace with the introduction of new technologies. Some organizations, like the World Instrumental Transcommunication (ITC) group formed in Luxembourg in 1985, have repeatedly followed Raudive's lead in using technological analogies to legitimize the existence of the spirit world (Raudive had referred previously to a 'Radio Peter'—which was a theoretical radio station translating the communications between the physical plane of congealed energy and the ethereal realm of pure consciousness). Likewise, the ITC have gone as far as to suggest that spirit entities use their own 'equipment' to interface with our own, or—taking this line of reasoning a step further—claiming that they are either supercomputers or (in the same vein as Bateson's earlier comments on noise) a kind of immaterial *Ur*-form or 'ground' that allows for the creation of more distinct, individual structures or 'figures.' This is interesting to ponder in light of other observations made earlier about 'figure and ground,' particularly when considering that the EVP listening experience itself relies so heavily on the perception of a kind of audio bistability.

---

[75] Alcock, "Electronic Voice Phenomena."

The ITC treat EVP as only one of a variety of methods by which interdimensional communication may be taking place: their explorations extend to the capture of apparitional images on television and videotape, informed by the work of Klaus Schreiber and Martin Wenzel and their development of a Vidicom device whose usage is explained as follows:

> As with the original radio EVP, the phenomenon could only be revealed by slowly scanning through a videotape recording (to find the single paranormal frame amongst the 50 interweaved frames-per-second). From the interesting frames Schreiber made still shots. Images that had good picture quality and were promising he filmed again on a second videotape using the same procedures... Schreiber noticed how recognisable faces would 'build up' out of washed-out images, as if during this 'birth' procedure a form-shaping 'morphogenetic' field may be active which strives for maximum similarity to a matrix picture... this could compare to the shaping of a life form out of ectoplasm during spirit materialization.[76]

Again, the activities described here are uncannily adjacent to contemporaneous ones occurring outside the realm of time-based audiovisual technologies, such as one unsettling experiment with a Polaroid still photomontage carried out by Ian Sommerville[77] at the famed Parisian 'Beat Hotel' frequented by key members of the Beat poetry movement.[78] Incidentally, the popularization of EVP techniques also owes itself in part to the exploratory efforts of Sommerville's notable collaborator, the author William Burroughs, whose 'cut-up' methodology and iconoclastic attitude towards given cultural templates had already energized a good deal of the 'post-industrial' music and sound arts communities. For this community, whose aesthetics often implied a sort of ghostly residual memory imprinted upon emotionally resonant sites such as factories and concentration camps, EVP would likely have found its way to their recordings without the endorsement of a tutelary figure such as Burroughs (though plenty of extant documentary evidence confirms the author to be an influence in this regard, i.e., the influential industrial music group Throbbing Gristle,[79] while detailing their creative methodology on their documentary album

---

[76] Lander, *Beyond the Dial*.

[77] Ian Sommerville (1941–1976), a British mathematician and computer programmer, was part of the 'Beat Hotel' circle with William Burroughs and Bryon Gysin, with whom he developed the stroboscopic 'dreamachine' project.

[78] "Another fascinating project that preoccupied Ian was using a newly developed Polaroid Model 800 instant camera to take a shot every few minutes for all his waking hours, every day. At the end of each day he stuck up on a wall all the photographs he'd taken that day. He then took a Polaroid photo of that array, each photo butting up to its neighbor in a strict tiled pattern. He repeated this on a daily basis until he had enough of the second-generation photos to make an array of them, which he then re-photographed... Starting to emerge out of the matrix was the unmistakable impression of a human form, a ghostly image that appeared to hover in front of the montage." Wyllie and Parfrey, *Love, Sex, Fear, Death*.

[79] Throbbing Gristle were an English music and visual arts group, officially formed in 1975 widely regarded as pioneers of industrial music. See Ford, *Wreckers of Civilisation*. See also Bailey, *Micro Bionic Radical Electronic Music & Sound Art in the 21st Century*.

Grief, cite Burroughs's interest in the subject as a direct influence). Elsewhere, a 7-inch single released on paranormal researcher and studio-artist Michael Esposito's Phantom Plastics record label, featuring Esposito in collaboration with Swedish artist Carl Michael von Hausswolff,[80] is comprised of EVP recordings done at the Mexico City site where Burroughs accidentally shot and killed his common-law wife Joan Vollmer in an incident that plagued the author for the remainder of his life.

At any rate, Burroughs's combative approach towards re-editing reality with one's own montage work also informed his approach towards the apophenia-fueled phenomenon of EVP (and inadvertently confirmed foregoing statements about the non-passive nature of perception). Burroughs's claim that "you control what you put into your montages; but you don't control what comes out," implied that individuals working with this medium might, via 'top-down' processing, choose what unique kernels of sonic information to focus on, but left the process of constructing meaning open to other forces less clearly defined or understood. Upon reading Raudive's *Breakthrough*, Burroughs incorporated similar insights into an essay titled *It Belongs to the Cucumbers*, which was first presented at the Naropa Institute at the end of the 1970s and later incorporated into an essay collection, *The Adding Machine*. Despite its relative age, Burroughs's assessment of EVP in this essay is remarkably similar to the type of impressions received by modern-day enthusiasts of the same:

> The speech is almost double the usual speed, and the sound is pulsed in rhythms like poetry or chanting. These voices are in a number of accents and languages, often quite ungrammatical... "You I friends. Where stay?" sounds like a Tangier hustler. Reading through the sample voices in *Breakthrough*, I was struck by many instances of a distinctive style reminiscent of schizophrenic speech, certain dream utterances, some of the cut-ups and delirium voices like the last words of Dutch Schultz.[81] Many of the voices allegedly come from the dead. Hitler, Nietzsche, Goethe, Jesus Christ, anybody who is anybody is there, many of them having undergone a marked deterioration of their mental and artistic faculties. Goethe isn't what he used to be. Hitler had a bigger and better mouth when he was alive.[82]

Burroughs's even-handed approach to the subject makes good on the assessment of researcher Deborah Dixon, who likely speaks for a good deal of the EVP enthusiast community when saying that "EVP research tells us much about the authoritative status of cause and effect explanatory frameworks, as well as the innocence accorded

---

[80] Carl Michael von Hausswolff (1956–) is a Swedish artist who has investigated the perceptual limits of various sensory phenomena, while also working as a conceptual artist, filmmaker, composer, and curator.

[81] Dutch Schultz (born Arthur Simon Flegenheimer, 1901–1935) was a New York City-based mobster who came to prominence via illegal gambling and bootlegging operations; the 'last words' mentioned by Burroughs were bizarre stream-of-consciousness musings (e.g., "a boy had never wept nor dashed a thousand kin") uttered during police questioning of the dying Schultz on his hospital bed.

[82] Burroughs, *The Adding Machine*.

technological apparatus."[83] Dixon's defense of EVP on the grounds that it "resists a simple categorization of practices and beliefs as either 'the occult' or 'science'"—an opposition she believes has been "used to valorize the latter"—is indeed applicable to a much broader range of investigations into phantasmatic phenomena, and forms part of a deeper historical narrative than may be immediately assumed. For example, Dixon is careful to remind readers of the spiritualist allegiances that were embraced by notable inventors, for example, Thomas Edison, while also noting how Nikola Tesla "speculated that the signals he began to pick up on his radio tower in Colorado in 1898 were the greeting of one planet to another" (interestingly, this potential source of EVP artifacts was actually rejected by Raudive, who claimed the utterances he captured were "too banal" to be interplanetary communications, inspiring Burroughs's retort that there is "no reason to believe [humans] have a monopoly on banality").[84]

Though some are far more enthusiastic about EVP's creative deployment than others, it could be said that the pervasive ambiguity accompanying the practice (both the ambiguity surrounding its origins, and the semantic ambiguity of the recorded material itself) is the very reason for its growing success as a sub-species of what we have called here 'the sonic blotscape.' This ties into an earlier cited interest in rescuing audible distortion from its status as "a flawed departure from a perfect copy and/or an unintentional byproduct of some faulty process or malfunctioning equipment."[85] In the case of EVP, this is paradoxical, since the embrace of recording conditions dreaded by professional sound engineers is at once a refusal of 'inhuman' digital communication, and an attempt to communicate with entities that are inhuman in another meaning of the word. Touching again upon apophenia, sonic researcher and artist Alex Keller provides one rationalization for why the communicative ambiguity of EVP attracts us, and for why EVP artifacts are associated with spirit communication:

> Your ear has a row of cilia, and each one represents more or less a specific frequency... When you get tinnitus, you've damaged the cilia in that [particular frequency range], and that damage creates that mechanical pressure and pain related to hearing that particular frequency, but I think that cognitively when you listen to something for so long, you perceive the sound as louder in general but you will also tend to reject things in that particular range. There's that weird thing where, when you're listening to static, sometimes you'll hear voices, because there's just so much stuff happening, there might be random things happening that are in the [audible] speech range. Our ears are tuned into the speech range, and that's why we tend to hear 'ghost voices' in static, because that's where our ears are most sensitive.[86]

---

[83] Dixon, "I Hear Dead People."
[84] Burroughs, *The Adding Machine*.
[85] Poss, "Distortion is Truth."
[86] In conversation with Thomas B. W. Bailey, November 30, 2018.

Despite the optimistic attitude projected by the founding fathers Jürgenson and Raudive, it has to be said that there is a certain impression many receive of these noisy, lo-fi artifacts, which can come about without any priming as to what the content is meant to be, and which closely mirrors the unvarnished assessment of sound artist and EVP-enthusiast Joe Banks: "the ambience of the recordings is menacing."[87] Accordingly, EVP has been a special point of interest for individuals interested in the 'darker' side of the phantasmatic; musicians in the so-called 'dark ambient' genre particularly seizing upon its ability to inculcate a feeling of uncanniness redolent of H. P. Lovecraft[88] or other notable vendors of cosmic and interdimensional horror. This is brought chillingly to the fore by many of the assumptions made about the 'speakers' of EVP (i.e., the characteristics that such voices would have when not being produced by human vocal cords and other fleshy apparatus); they are like modernizations of assumptions made in classic works on demonology and the occult, such as the claim in the *Malleus Maleficarum* that "when [devils] wish to express their meaning, then, by some disturbance of the air included in their assumed body, not air breathed in and out as in the case of men, they produce, not voices, but sounds which have some likeness to voices, and send them articulately through the outside air to the ears of the hearer."[89]

On a less fantastic level, much of the post-industrial music community has imbrications with dark ambience owing to their shared skepticism of authority and, moreover, the suspicion that official power structures such as the CIA are informed by an esoteric knowledge that is nevertheless still in play for any radical elements who wish to use it against the institutions. Lastly, the technical and procedural aspects of EVP contain parallels with certain tendencies within late twentieth- or twenty-first-century sound art. The practice of listening to extended periods of silence for a granule or two of meaningful communication has been a significant feature of the so-called 'lowercase' music genre, generally associated with artists such as Bernhard Günter and evident on releases like *Un Peu de Neige Salie* ('A Little Bit of Dirty Snow') and *Details Agrandis* ('Enlarged Details'). This was a tendency also given a significant boost by Francisco López and other artists who, though not exclusively absorbed in this technique, felt it wise not to neglect its emotional or intellectual potential within a more wide-ranging acoustic oeuvre.[90] In a strange parallel to Keller's foregoing explanations about sifting through a surplus of indistinct audible data to find meaning, Günter's more influential work played upon listeners' desire to find meaning within a *deficit* of audible information, to again latch onto 'something rather than nothing,' even in cases where 'nothing' is closer

---

[87] Banks, "Rorschach Audio."
[88] Howard Phillips Lovecraft (1890–1937) is today one of the most canonical writers of supernatural/horror fiction, with his interdimensional "Chthulu Mythos" inspiring entire sub-genres of the form as well as real-life devotees of the fictional entities presented within.
[89] Smith, *Muses, Madmen and Prophets*.
[90] E.g., in particular, López albums, *Warszawa Restaurant, untitled #74, untitled #91, untitled #150*, and the twenty-year compilation *Presque Tout*.

to the perceivable reality that exists prior to the emergence of phantasmatic effects. The strength of this music lies the way it reveals how lack of sound can be informative, and how a 'ground' of silence can transform what we would normally interpret as a kind of non-semantic, cosmetic enhancement (an isolated metallic glimmer, an arcing sine wave) into a 'figure' that becomes a peculiar carrier of meaning. Parallels within the larger art world could be drawn to artworks such as James Turrell's[91] simulations of Ganzfeld states (e.g., *Wide Out*, 1998); wherein loss of depth perception stands in for loss of audible information.

Though prima facie seeming to have little in common with a phenomenon like EVP, both that genre of audio information and the 'lowercase' style of sonic art have commonalities that go beyond their aesthetic similarities. Namely, they both rely upon a shift from passive to active listenership (or perhaps it would be better in light of the previous discussion about Auditory Scene Analysis to refer to this as *consciously* active listenership). Both attract listeners to participate in such active listening by holding out the promise of personalized meaning, of highly individuated semantic worlds, secreted within either the noise or the silence. Meanwhile, the standard EVP practice of *welcoming* degradations in audio fidelity, insofar as they provide a more fertile territory in which voices can manifest themselves, is very much in keeping with a larger embrace of *unwanted* (or what theorist Michel Serres might call 'parasitical') elements such as electronic feedback as the essential building blocks of a composition. These artists' deliberate inclusion of the 'white' surface noise containing all possible audible frequencies harks back again to Burroughs's encouraging creators to work with the whole spectrum of possible experiential impressions in order to craft a more authentically individualized reality, and at the very least there is an implicit symbolism of "diving beneath the surface [noise]" of mediated reality to more critically engage with life.

Given all that has just been said (and will certainly be said again) about EVP's representing a form of direct 'spirit communication,' it is perhaps best to bypass this debate completely and focus for now on something that can be agreed upon by both committed paranormalists and the more 'skeptical' sonic researchers who concur with what Banks expresses when he states "an understanding of the processes involved in EVP may also contribute to theories of art criticism, both in general terms and with particular reference to music and sound art."[92] Another of his realizations—that "the fog of noise that degrades these signals still seduces some people into suspending disbelief"[93]—also opens towards a much more in-depth discussion on the nature of the phantasmatic. In particular, it elicits a larger discussion on how the imprint of 'authenticity' that comes from the use of less-than-ideal technologies often helps to amplify our already strong inclinations towards apophenia (and the emergence of the

---

[91] James Turrell (1943–) is an installation artist with an intense focus upon light and its relationship to space, with many of his works creating memorable sensory and proprioceptive illusions.
[92] Banks, "Rorschach Audio."
[93] Ibid.

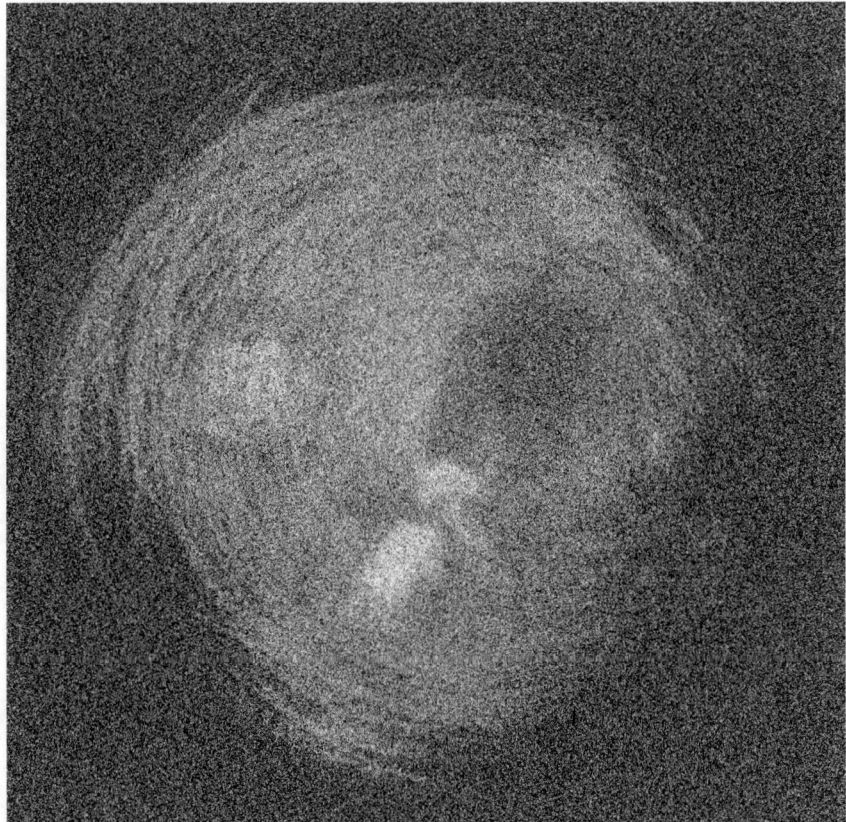

Figure 4.7 Raising the noise floor. Untitled drawing-photograph composite using a *Drawing Phantoms* graphic result and a found old Victorian photograph. Image by Barbara Ellison, 2019.

so-called 'hauntological' genre of electronic music, often using indistinct or reduced bandwidth elements to deal thematically with ghostly unrealized versions of the future, also points to this).

Notwithstanding individual beliefs and personal interests, it is important to stress here that our comprehension of this phenomenon has not diminished at all its poetic, emotional, dramatic, and artistic character. That is why one of the last versions of the *Drawing Phantoms* project was presented in performance under the title *Drawing EVP*.[94] Out of emotional and artistic respect for EVP, as well as a desire to potentially synthesize some of the conclusions from my other projects dealing with sonic phantoms, the source material for this particular compositional series would be more properly classified by the proposed variant OVP. This term, and the concept

---

[94] The program "Lost Spectres and Ethereal Voices," included my performances of *Drawing EVP* and *The Other Voice* at Störung 9 Festival, Barcelona, 2014.

it embodies, refers to what can be clearly envisioned as a phantasmatic phenomenon more wide-ranging in its intentions and in its methods of elicitation (when compared with the purely 'electronic' elicitation of EVP); one that, somehow in an animist fashion, encompasses all imaginable objects and the myriad unexplored ways of potentially interacting with them.

The *Drawing EVP* performance involves deliberately raising the noise floor (i.e., the measure of the signal that consists of the total amount of noise and the generally undesirable artifacts) of the contact microphones, and then applying extreme specific equalization-filtering techniques on this noisy matrix. It is typically one of the primary tasks of the sound engineer to banish these artifacts from radio communication, electrical devices, electromagnetic sources, cellular networks, and other consumer electronics, so as to keep the signal clear and uncompromised. As mentioned before, such noise fields can present interesting opportunities for discovering rich sonic sources: they can provide the starting point for fresh compositions and performances, as it were, rather than being the final step to be confronted before a composition or performance is considered 'audience-worthy.' If properly manipulated during the live performance, and taking into account the acoustic specificities of the space, these kinds of balancing and equalization techniques can eventually make the drawing sounds appear strikingly speech-like, and therefore generate a multiplicity of voice sonic phantoms. With practice, it is actually feasible to 'sculpt' voices out of the background noise field by effectively simulating different vowel and consonant sounds through manipulation of audio filters at the appropriate frequencies.

Perhaps more on the distinction between EVP and OVP needs to be said before this section is properly concluded. We have seen here that, as it stands in its own one-century-old tradition, EVP manifestly belongs to a profuse lineage of techno-mystical cultural constructions—such as spiritualist photography or electrical resurrection—that took shape as a consequence of the widespread social experience of new technologies. As evinced by devices such as Spiricom,[95] the technological term 'medium' can maintain the ethereal or sidereal quality and powers of interdimensional understanding once attributed to *human* 'mediums,' acknowledging in the process a kind of sentience perceived in the technological extensions of human communication. This was not always the case: at one stage in the development of modern paranormal research, the point was to look towards capacities for human communication that had not yet been captured by science, rather than to harness new technologies towards that same end. In the paranormal classic *Phantasms of the Living* (largely credited to F. W. H. Myers in 1883, but also co-authored by Frank Podmore and Edmund Gurney) humans were encouraged to "use extrasensory perception—super-psi, as it is called in the literature of parapsychology—to acquire information that had never been in any mind."[96] Nevertheless, the prevailing attitude of the Society for Psychical Research

---

[95] George Meek and Bill O'Neil between 1979 and 1982 developed a device they named the 'Spiricom' (short for spirit communication). The Spiricom was a set of thirteen tone generators spanning the frequency range of the adult male voice, invented for two-way communication with spirits.

[96] Quoted in Gray, *The Immortalization Commission*.

was that the methodology and standardization of science could be applied to the psychic realm, an approach that has enlivened a techno-gnostic approach to reality and creativity, while simultaneously stoking fears about, for example, military applications of 'remote viewing' psi techniques.

Perhaps with a relatively unconscious will to debunk this privileged techno-gnostic status of the 'electronic,' or more likely to exist alongside it in a complementary fashion, 'Object Voice Phenomena' thus expresses the claim that all objects, and not only the 'electronic,' are potential sources of 'voice phenomena.' They all potentially contain those presences: if and when properly activated—as when drawing on amplified surfaces—they will reveal an endless multiplicity of voices. Electricity is not the only medium. Any object, instrumental or not, even the most apparently quiescent, is also a medium. We just need to know how to induce the phantasmatic in it.

Figure 5.1 Iqaluit, Baffin Island, Nunavut, Canada. Landscape just outside the town and young *katajjaq* performers, Sandra Ikkidluak and Crystal Mullin, during a recording session at the Inuksuk High School. Photographs by Barbara Ellison, 2010.

5

Phantasma Humana: The Realm of the Voice
Compositional Series: *Vocal Phantoms*

5.1 *Katajjaq*

In December 2009, in the middle of the Arctic winter, I travelled with fellow artists Stephanie Pan and Stelios Manousakis to the community of Iqaluit on Baffin Island (Nunavut, northern Canada), primarily to have a direct experience of *katajjaq*, or Inuit throat singing,[1] which is uniquely found among the Inuit. The traditional form of *katajjaq* consists of two women facing each other standing very close, holding each other's arms whilst they rapidly exchange back and forth an alternating sequence of guttural and breathy sounds and noises, in the process creating an uninterrupted phantasmatic sonic stream. With the invaluable assistance and outstanding hospitality from Vinnie Karetak, as well as the kind collaboration of many people in the Iqaluit community (see acknowledgments), we had the joy and privilege of experiencing the remarkable *katajjaq* first-hand, being allowed to carry out audio and video recordings, and receiving inestimable direct insights into the techniques and history of *katajjaq* from local throat singers who were extremely generous and playful in sharing their knowledge and experience. This experience was to later inspire and kickstart an entire series of compositions based, in a more general sense, on interlocking techniques for both real and synthesized voices. These were developed under the generic title of *Vocal Phantoms*.

Katajjaq is not to be confused with the more popularly recognized Tuvan or Mongolian overtone throat singing (a.k.a. *khoomei* and *khargyraa*), which is distinguished by the production of two or more notes simultaneously by one singer. Having been embraced by adventurous Western ears thanks to the efforts of enigmatic stars such as Sainkho Namtchylak,[2] and given a stamp of cultural legitimacy by UNESCO in 2009 (i.e., being placed upon that organization's Representative List of the

---

Readers can listen to several pieces of this compositional series at the website www.sonicphantoms.com

[1] The name for vocal games in Canada varies with geography: *katajjaq* or *katadjak* in Nunavik and South Baffin, *Iirngaaq* in some Nunuvut communities, *Piqqusiraarniq* or *Pirkusirtuk* in Igloolik and Baffin Island, *Qiarvaaqtuq* in Arviat, and *Nipaquhiit* in some Nunavut communities.

[2] Sainkho Namtchylak (1957–) is a Tuvan-born vocalist who was part of the state's folk music ensemble as well as more characteristically 'experimental' groups; she is one of the most successful female practitioners of overtone singing.

Intangible Cultural Heritage of Humanity), the latter is the form commonly referred to or imagined when the topic of throat singing is introduced. In addition, a variety of throat-singing techniques are practiced worldwide, such as in Sardinia (*cantu a tenòre* from the Barbagia region), Tibet (Buddhist chanting), the Balochi *Nur Sur* (an ancient form of throat singing still popular in Afghanistan, Iran, and Pakistan), and the *Umngqokolo* traditional overtone singing of the Xhosa in South Africa. More closely related to *katajjaq* in terms of technique and sounding image is *Rekukhara*, a vocal game once known to be widely practiced amongst the Sakhalin Ainu people of Russia.[3] Vocal productions with the so-called 'panting style' have also been found in recordings of *Pič Eynen* of the Chukchi communities of Siberia.[4] French-born Canadian musicologist and semiologist Jean-Jacques Nattiez clarified that *katajjaq*, for those new to the form, is *not* the "main musical genre of the Inuit culture" (a distinction that he reserves for "the song of the drum dance"),[5] even though *katajjaq* and related forms can be performed at the same ceremonial event with drum dances.

Katajjaq is an important part of Inuit culture and, unlike other forms of throat singing wherein men are traditionally the performers, Inuit throat singing is practiced almost exclusively by women. It was traditionally used to sing babies to sleep or as a form of entertainment in games that women played during the long winter nights while the men were away hunting, sometimes for periods of months: a context that is important in understanding the technique and the form of this practice. Aside from the personal and collective entertainment aspect of the game, it is also used for practical purposes, such as a breathing exercise to prepare for bad weather. Naturally, the aforementioned association of the form with mothering duties is another key difference between it and Tuvan throat singing, as the latter has traditionally been the domain of male herders and has only recently in its history been performed by female vocalists. It is therefore a 'game' in a very rich and wide sense of the term (particularly, as belonging to the types of games known as *pileojartuq*, or two-person contests), while also being a practice that would include very relevant cultural roles having to do with bonding and communication within the community.

Nattiez has declared this form to be a "multifunctional host-structure,"[6] and the evidence of that is clear in the considerable number of functional roles the game can play, as well as the spontaneity involved in initiating games (*katajjaq* sessions often involve competing 'teams' whose members face off against one another in pairs until a sufficient number have been eliminated from one side). Besides the aforementioned role of entertainment among the women of the community while their husbands were occupied with hunting, it is additionally used as well as a ritualistic means of intercession to either bring the hunters home safely or to provide them a favorable

---

[3] According to an interview with the daughter of the last practitioner of this style (who died in 1973), the *rekukhara* was often done during the *iomante* ritual, the slaughter of a brown bear, as the produced sounds from this game would symbolically refer to the cries of the bear. Nattiez, "Inuit Throat-Games and Siberian Throat Singing."
[4] See Nattiez, *Some Aspects of Inuit Vocal Games*; Beaudry, "Singing, Laughing, and Playing."
[5] Nattiez, "Inuit Throat-Games and Siberian Throat Singing."
[6] Ibid.

outcome for their hunts, occasionally by influencing the elements themselves. It is also employed to 'fast track' the return of geese to the region and, by extension, the coming of spring. Perhaps because of the pagan affinities suggested by some of these ritual functions, the form was banned in the 1920s by Anglican missionaries. Missionaries from other denominations eventually encouraged its continuation decades later, and thankfully it is experiencing a more recent contemporary revival. *Katajjaq* today has become a means of preserving and promoting Inuit cultural pride, heritage, and tradition and this holds especially true among younger generations, who believe that learning it from their elders connects them to and strengthens their bonds with Inuit culture. A rough outline of the form's technical and stylistic peculiarities is, of course, in order. *Katajjaq* is usually played with two women, but can involve up to four or five women at a time.[7] The most typical practice of the *katajjaq* has been discussed at the beginning of this section, though earlier iterations of the game featured a style of play in which the women's mouths were almost touching, in order to better use the opponent's mouth as a resonating cavity. This is apparently no longer done.

The technique involves making short, sharp, rhythmic inhalations and exhalations of breath, and contrasting currents of low and high tones (the lower tones produce what has been popularly recognized as the 'throat' sound, though this is actually absent from some regional variants of the form). One woman sings or vocalizes the main rhythm pattern with silent gaps, so the second woman can respond in the gaps with the same rhythmic pattern, with a variation of it or with some other complementary rhythm melodic pattern.

Though, again, some regional variants exist that feature vocalization made from strung-together sentences, a narrative quality is not essential to the final result. In most cases, sound is made continuously from single morphemes, using sounds that are voiced or unvoiced, both on inhalation and exhalation. This results in something comparable to hyperventilation or breathlessness, though perhaps with more of a modicum of control. As the game progresses, the singers try to fuse the combined sounds they produce so that the 'audience' cannot distinguish one singer from another. When one of them introduces a new sonic element into their pattern, the other has to follow immediately to match. They follow each other in interlocking mode so that the strong accent of one partner's vocal pattern is superimposed on the weak accent of the other (an effect Nattiez refers to by the more technical term 'crossdephasing').[8] The *katajjaq* is a very inventive game of playful endurance, and if listeners do not initially understand the ludic character of the singing, this becomes eminently clear when it ends with the laughter of a defeated party (who can also become recognized as the 'loser' of the game when she cannot keep up with the pattern matching or when she runs out of breath).

This superimposition and alternation of weak and strong accents creates perceptual ambiguity, which is an essential part of the form. Using this technique, the singers deliberately create a powerful illusion in which the listener cannot determine which voice is singing which part: in the *katajjaq*, all sounds seamlessly appear to come from

---

[7] Nattiez, "Inuit Throat-Games and Siberian Throat Singing."
[8] Nattiez, *Some Aspects of Inuit Vocal Games*.

one source. Through this ambiguity one experiences an impressive and seamless stream of sounds and noises, which generate a multitude of interesting sonic phantoms. Aside from the playfulness of the game, this illusion is one of the most fascinating aspects of this practice, as there is a profound perceptual mismatch between what is seen and what is heard. In this sense, it is a fascinating object of study for those already familiar with the 'sensory jamming' illusion known as the McGurk effect,[9] in which conflicting audial and visual feedback (e.g., a mouth appearing to speak the phoneme 'ga' while the sound 'ba' appears to be spoken during lip reading) will cause the perception of an 'intermediate' feedback, namely a third unique phoneme that is neither of these.

Returning briefly to the topic of other throat-singing variants, some interesting stylistic similarities have been identified between styles originating in geographic areas thousands of miles apart from one another. *Katajjaq*, as mentioned above, has been likened to other throat-rasping duets such as *rekkukara*, traceable to the indigenous Kraft Ainu people originally from Sakhalin Island[10] (a form that has almost disappeared). These throat games, which also seemed to be exclusively practiced by female vocalists, have been described by Nattiez as being "constructed on short, reiterated motifs," with "not more than three or four levels of intonation, which are organized in patterns." Nattiez contends that their most intriguing shared feature is the shared "sound materials" themselves, particularly the "combination alternating between voiced and unvoiced sounds, inhaled and exhaled."[11] *Rekkukara* nevertheless has some features worth examining on their own merits. For one, it is more purely devotional than *katajjaq*, given its usage during the bear festival central to the animist beliefs of the Ainu (more specifically, during the so-called 'rejoicing' portion that forms the prelude to a bear's ritual sacrifice). In fact, the usage of the throat technique itself is done in emulation or mimicry of a bear's cry. Beyond this, though, are some performance peculiarities to discuss. One of these is the speculated use of hands during the game as both a theatrical element *and* acoustic element (a 'megaphone' in Nattiez's reckoning) that might serve to merge two partners' sounds into a single stream, as it were. Both partners cover their mouths with their hands, which touch each other, creating a kind of resonance box for the sounds.[12]

Elsewhere, the aforementioned *Pič Eynen* form, literally translating to 'sing with the throat' and distributed among numerous Siberian tribes, expands the number of singing participants: rather than exclusively using a duo configuration, *Pič Eynen* features a leading voice and a number of other singers who improvise in turn. With this added versatility, it also seems to be more clearly adaptable to situations in which a ritual dance is occurring, unlike *katajjaq*, which is performed independently of

---

[9] MacDonald and McGurk, *Visual Influences on Speech Perception Processes*. Also see Macknik, Martinez-Conde, and Blakeslee, *Sleights of Mind*.

[10] The Ainu from Sakhalin Island were relocated to Hokkaido following its annexation by the USSR after the Second World War. According to Nattiez, the last immigrant from Sakhalin to have performed the *rekkukara* in Hokkaido died in 1973. Nattiez, *The Rekkukara of the Ainu (Japan) and the katajjaq of the Inuit (Canada)*.

[11] Ibid.

[12] Ibid.

such activities. In all known variants, however, the presence of sonic phantoms acts as a common feature transcending the varying levels of spiritual significance or social utility accorded to the singing technique.

5.2 *Vocal Phantoms #18* (for Live Voices)

Vocal Phantoms #18 is a work I composed for an unspecified number of voices, inspired by techniques employed by throat games such as the *katajjaq*, that can be adapted to work for multiple voices (this will obviously affect the textural complexity of the piece). Although the number of voices is variable, it is nonetheless desirable that each set of two partners has voices that match each other's timbre as closely as possible. The version described here is for four voices, divided into two groups of two interlocking duos.

The basis for this piece is a series of brief, alternating, and looped vocal sound patterns arriving in pairs. These sounding patterns are composed of selected vocables or morphemes (i.e., non-lexical combinations of certain sounds or written characters that have no discernible meaning) that are vocalized as discrete units, such as *em-ha*, *ha-mahe-ma*, *hi-yah*. Each one of these small units, essentially separated by breathing intervals, is what I term a 'motif' (*pace* Nattiez), which comprises the atomic elements of the piece. Each motif gives rise to its own spectral and temporal patterns, which upon repetition bestow it with its distinctive character. For the phantom-generating process to be truly effective, the sounds contained in each motif must be generated both by intake and expulsion of the airstream. This then results in a pattern of alternating breath, with corresponding voiced and unvoiced sounds (unclearly pitched)[13] producing distinct intonation and noise patterns. The act of breathing should ideally be as audible as possible, and can be used in a melodic way. Each respective motif demands different combinations of a kind of half-whispered and half-sung vocalization. The succession of syllables or morphemes chosen, which make up the motif, gives rise to a constantly repeated rhythm, which creates a regular intonation contour (albeit not necessarily one with fixed pitches).

To generate the sonic building blocks of this composition, I thoroughly considered and explored all of the possible motifs available and compiled a repository from an adequate number of them. From one motif to the next, there were differing rhythmic and morpheme patterns, intonation contours, voiced or voiceless (half-whispered) patterns, each with different use of inhaled and exhaled breathing sounds. This serves to provide a significant openness in the composition, since I can then select, combine, and adapt diverse motifs from this repository depending on what I feel can work best for a specific performance. I can thus modify the program at will in accordance with the unique players and set duration. Regardless of the motif's features, each one of them is repeated a specified number of times between the performing partners, which establishes the stable ground pattern and creates the illusion of a continuous stream of

---

[13] Voiced pitch has a periodic fundamental frequency, or is clear sounding pitch as opposed to voiceless pitch, which has no clear sounding fundamental.

sound. The partners' motifs are either identical or a variant but are predominantly offset in time so as to create a gradual de-phasing between the vocal layers. The total repetition count of each motif will depend on the overall duration of the version of the piece.

There are variants on the types of structural organization used to combine the motifs. In the score for this piece, the motif patterns of voiced and voiceless sounds, inspirated and expirated, are all differentiated with various graphic notation symbols.[14]

The structural organization of all these materials is based on three types of alternation:

1. The two players of each pair alternate a single, repeated motif 'A.'
2. The two players of each pair alternate two different motifs 'A' and 'B' (occasionally, they might switch to a unison configuration of the motifs).
3. A series can be created in which the two voices alternate in performing motif 'A,' then motif 'B,' then motif 'C,' and so forth.

The polyphonic approach to the organization and performance of this piece gives rise naturally to the same phenomenon that is patently utilized in the *katajjaq*: a striking mismatch between the motor and the sounding image; the difference between what is played and what is heard.[15] Regardless of which particular set of interlocking parts is put into play to construct a specific version of the composition, the listeners' perception is fundamentally guided by the auditory streaming effect and they will hear the emergent Gestalt in the place of constituent parts. Due to the use of very tight alternation between parts and their pitch and timbral organization, a special sort of perceptual confusion arises. Namely, listeners perceive patterns that are not played as such by any of the singers, but are instead results of the perceptual restructuring mechanisms of the brain. To elaborate upon the mismatch phenomenon already mentioned, this perceptual confusion is enabled by the seeming disconnection between the regular and symmetrical structure of the music's formal components, and the asymmetrical shape of the emerging phantom patterns.

When all of these compositional strategies are woven together, a sort of secondary composition takes place in which ambiguity and confusion themselves become the compositional materials, so to speak: the contrast between what is being played, versus what is being heard, initiates the transition from passive to active listening that has been a common thread through this text. However, as natural and simple as this form of transition sounds when described, this is not a process to be entered into without a lack of rigor or with an attitude of deliberation: achieving interesting results in this territory of intentional confusion requires a careful, disciplined organization of the elementary units, so that, despite their patent individual simplicity, they give rise to a multiplicity of rich auditory streams for the listener. This allows, for example, a simple isolated pattern to be organized so as to give rise

---

[14] In the score I use a notation I have adapted from a combination of Beaudry and Nattiez's detailed transcriptions and analysis of the *katajjaq*. Beaudry, *Toward Transcription and Analysis of Inuit Throat-Games*.

[15] This can be seen as a very intriguing side effect of the process of auditory stream segregation that takes place in music that exploits perceptual ambiguity.

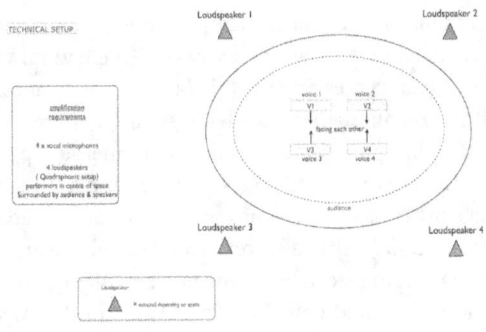

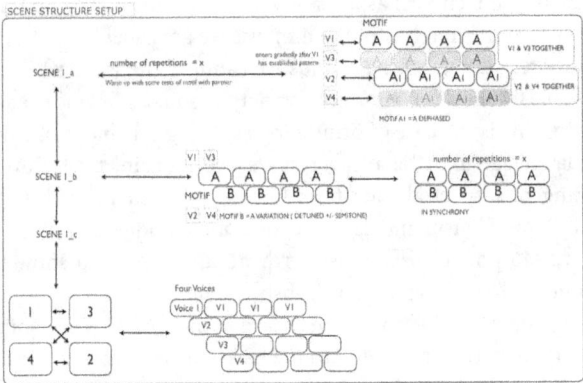

Figure 5.2 Section of the score/diagram for live performance set-up of the piece *Vocal Phantoms #18*, showing examples of organization of motifs and scenes in the piece. V1, V2, V3, V4: voices of different singers. A, B: motifs, in temporal sequence from left to right.

to complex sub-stream patterns, or for an individual performer to generate a kind of 'indirect polyphony.'

5.3 Ancient and Recent Vocal Techniques of Illusion

> *There are many vocal sounds of which it is not immediately obvious how they are produced (for instance, on an inhalation or an exhalation) or obvious through which organ in the tract—the uvula, the tongue or the larynx. These phenomena are even mysterious for colleague sound poets-experts, you would think; we regularly exchange questions on this.*
>
> —Jaap Blonk[16]

---

[16] Jaap Blonk (1953–) is a Dutch vocalist/sound poet who has been instrumental in the use of electronic vocal modification, while also being a regular performer of "classic" sound poets' texts. See Blonk and van Peer, "Sounding the Outer Limits."

The *katajjaq* technique outlined above is but one in a vast array of vocal techniques capable of producing illusionistic phantom results; and it is also just one form that gains much of its appeal from its relative closeness to the natural world. The sounds of nature have been in many cases a powerful source of inspiration in the music, as well as in other cultural manifestations involving sound, of different cultures around the world, with vocal techniques of mimesis having numerous applications that, like *katajjaq* itself, are not limited to aesthetic ones (the usefulness of such skills among our hunter-gatherer ancestors probably does not need to be elaborated upon too much). Nevertheless, some of the more artistically conceived examples of illusionistic technique strongly affirm how challenging it can be for the perceptual process of prediction error minimization to be foolproof, and how the imperfection of that mechanism can lead to creative and cultural evolution.

Speaking about the technical aspects of vocal illusionism, Trevor Wishart affirms that "the sky is the limit" in terms of the mimesis we are capable of, stating that "if we do not make any restriction on the sounds we can use, the immense pliability of the voice makes it able to mimic an enormous variety of sounds."[17] One notable proof that Wishart gives for this is the case of ornithologists using syllables not commonly used in their native languages to better attract birds, as well as using the technique known as 'formant tracking' or a 'spectral glide' from low to high formants.[18] The fact that Dada 'phonetic poet' Raoul Hausmann also used such a technique in a non-ornithological context (in his 1946 poem *Birdlike*) is worth noting, if only to show the spread of specialist techniques to more ambiguous, abstract ends.

For the Inuit people, all kinds of sounds heard in their Arctic environment might be inspirational and are permitted for use as the basis for the voiced patterns in *katajjaq*. Nattiez has noted how the *katajjaq*'s basic rhythmic and breathing structure absorbs sound sources of various origins: meaningless syllables, archaic words, names of ancestors or old people, animal names, place names. It is often the case that *katajjaq* vocalizing also imitates animals, which in Nattiez's recordings alone included seals, walrus, geese, seagulls, dogs, and mosquitoes. Furthermore, they often imitate natural sounds like wind, water, or waves on the seashore. All told, the range of sounds used (musical tones, vocables, imitative natural or animal sounds, noises) is extensive. The Inuit are open to using all such sounds as long as they can they can maintain the illusion that all sounds come from a single source.[19]

In relation to the generation of sonic phantoms, we find again a crucial structural interlocking organization of the combined parts. The partners vocalize their motifs

---

[17] Wishart, *On Sonic Art*, 289.
[18] These interchangeable terms refer to a compositional concept in which the vowel quality of a vocal sound is modified. This concept begins from the premise that overtones, timbre, and spectra all contribute to perception of vowel quality, and demonstrates that all of these characteristics come into play when smoothly and seamlessly "gliding" from one vowel sound to another. For more information, see Erickson, *Sound Structure in Music*.
[19] Beaudry, *Toward Transcription and Analysis of Inuit Throat-Games*; Beaudry, "Singing, Laughing, and Playing."

in alternating patterns of voiced/un-voiced and inhaled/exhaled sounds, with the total effect resulting from the motivic superposition of the de-phased voices.[20] If, for example, each motif contains a low-pitched sound followed by a high-pitched one (or the contrary) and if the two voices follow each other by half a beat, we get the impression of hearing two distinct streams of low-pitched and high-pitched sounds, respectively. This streaming effect contributes to the puzzling conflict between the 'played' motor image and the experienced auditory image in the *katajjaq*, and hence to the intriguing sonic phantoms that result from the seamless interlocking of the two vocal parts. It is a sonic phantom that could be compared to the inherent patterns generated in the *amadinda* xylophone music in that both interlocking voices are creating each of the sounds heard. Just as intentionally, in the vocal games there is the conscious aim to reinforce this perceptual illusion. One voice is producing a sequence of hard accents (beats) and the other is producing a line of soft accents (off-beats)—resulting in the two different voices creating these two distinct streams of sounds. The stronger the illusion, the more successful the game is: namely, it is an intrinsic goal of the practice to make it difficult for the listener to tell who produces which sounds, even when one understands how the game is played. Factors that contribute to this illusory effect include spatial proximity of the players and similarity in their sounding vocal qualities (range, timbre). These qualities are deliberately and skillfully used by the players to encourage perceptual ambiguity in the listener, which results in the desired prominent mismatch between what is seen and what is heard. Auditory streaming processes, linked with these specific timbral, pitch, and spatial organizations, are responsible and can account for the appearance of sonic phantom patterns.[21] The more skilled the players in the execution of the *katajjaq*, the more spectacular the sonic phantoms will be.

Interestingly, some radically different cultural manifestations that have no ethnic, historical, or geographical connection whatsoever with the *katajjaq*, employ somewhat equivalent vocal techniques for similar illusory interlocking effects. It has been demonstrated elsewhere, for example, that the sounds of percussion instruments from numerous divergent cultures can be closely mimicked by plosive consonants and stops, which have sharp attack times similar to those of drums, and natural reverberations of a bass drum can also be approximated by phonemes ending with an 'm' consonant. There is indeed a globally distributed phenomenon of 'rhythm-speech' that includes, for example, the sharp attacks of the *gamelan mulut* or 'mouth gamelan' of Balinese vocal music, which has already been touched upon briefly, and merits returning to in this section, if for no other reason than the fact that the hocketing employed in this vocal style makes it more of an extension of the very similar percussive patterns that it regularly accompanies than a pure 'stand-in' for them. The eight-pitched bamboo *anklung*, a percussion instrument widely considered

---

[20] Beaudry, Toward Transcription and Analysis of Inuit Throat-Games; Beaudry, "Singing, Laughing, and Playing."
[21] Heise and Miller, "An Experimental Study of Auditory Patterns"; Miller, "The Trill Threshold".

as a predecessor to the gamelan, relies exclusively on hocketing for its unique effects (and, in an interesting bit of cultural parallelism noted by ethnomusicologist Victor Grauer, all such instruments have been used for Balinese funeral ceremonies, making them somewhat akin to Western church bells that also achieve their characteristic presence largely due to hocketing):

> I know of no other musical style with a distribution so strongly suggestive of great age. Found most commonly in Africa, the home of what may have been the first truly human creatures, this form of singing could well have been diffused with the earliest migrations of man. It would be an irony indeed if this style, so simple, really, yet so complex from the point of view of most musicologists, should turn out to be that wonderful mirage, shimmering and shining through the first paragraphs of many a history of music—man's first musical utterance. This would effectively reverse commonly accepted notions which see music beginning with grunts and groans, thenceforth evolving painfully towards polyphony. Early man was not, after all, quite so naive as some of our musicologists believe.[22]

Elsewhere, anthropological researcher Tok Thompson claims that this global distribution of 'rhythm speech' accounts for everything from "[Indian music], with its *bol* form" to "'Celtic' regions with [their] *puirt à beul*, or mouth music"[23] and onward to other traditions such as the 'scat' vocal accompaniment to jazz. *Bol* music, a musical style whose name derives from the Hindi *bolna* ['to speak'], is described by ethnomusicologist James Kippen and computer scientist Bernard Bel as "constitut[ing] an oral notation of onomatopoeic syllables that represent [tabla] drum strokes"[24] (the name was applied also to Kippen and Bel's 'Bol Processor,' a computer tool for musical composition and analysis that aimed at emulating polymetric features of Indian music).

The origins of the aforementioned *puirt à beul*, which had to do with providing rhythmic backing for dancing when instruments were unavailable, are similar in that sense to the origins of a much more recent and universally recognizable form of 'rhythm speech,' insofar as the latter also came about partially as a result of having limited financial resources to spend on electronic instruments. The form in question, briefly mentioned earlier, is called 'beatboxing.' It is part of a popular modern tradition of vocal percussion; a relatively recent urban practice that exploits many vocal techniques that, with or without the use of microphones, imitate the contemporaneous beat-producing machines, such as the now-classic Roland TR-808. This is somehow a paradox, since these machines in fact already imitate—at least in their origin—acoustic percussive sounds, and mark beatboxing as a 'folk art' containing the requisite qualities of 'authenticity' and improvising with the tools near at hand, while at the same time

---

[22] Grauer, *Some Song-Style Clusters*.
[23] Thompson, "Beatboxing, Mashups, and Cyborg Identity."
[24] Kippen and Bel, "Computers, Composition, and the Challenge of 'New Music' in Modern India."

aspiring symbolically towards the condition of the machine. The form dates to the early 1980s, with Doug E. Fresh (née Davis) as its likely originator (he has claimed for himself the title of 'Original Human Beatbox'), and with groups like the Fat Boys[25] cementing its popularity via music videos and the semi-autobiographical film *Krush Groove*. Despite the technique's seemingly diminished usage in the more successful hip-hop productions of the present century, it could be argued that the perceived virtuosity involved in beatboxing played its role in cementing the international popularity of hip-hop culture's 'Golden Age.'

The nature and innovation of performance techniques being explored in the beatboxing style are truly phantasmatic, and their distinctiveness and extremity also make them worth mentioning here. Within the beatboxing community, artists are actively and playfully pushing boundaries to virtuosic mastery above and beyond what one could imagine possible with the human voice. Beatboxers use an incredibly inventive array of extended vocal techniques along with 'close mic' techniques to explicitly modify the timbral characteristics of the voice. Like the *katajjaq*, non-syllabic patterns and inhaled sounds take center stage, as well as a rich, diverse repository of other techniques such as the use of trills, rolls, buzzes, and an astonishing array of vocal qualities (head voice, chest voice, breathy voice, growls, etc.).[26]

As in the case of the Inuit vocal games, beatboxing includes the widespread use of inhaled and exhaled sounds to produce very convincing sonic phantoms. In the case of beatboxers, however, they typically perform alone and their aim is precisely to create the illusion of a kind of polyphony with multiple 'rhythm-box' sound sources. In an article characterizing the beatboxing style, the authors suggest that these types of rapid alternations are used to create auditory illusions and are employed because they can encourage the perceptual timbral streaming of the sound layers.[27] They aim to trick the listening brain into perceiving certain sound events (basslines, drum machine imitations, vocals) as taking place simultaneously. Some of these techniques also involve the rapid alternation between vocal timbres, producing multiple streams. In fact, the inventory of such fascinating vocal effects is taken further still by Tok Thompson, who suggests that beatboxers also act as occasional emulators of the "sounds that resemble electronic sampling" or "the 'scratching' of a phonograph needle against a vinyl LP,"[28] and one does not have to dig too deeply into scene lore to find virtuoso exemplars such as Rahzel (née Rahzel Manely Brown)[29] whose vocal intricacies allow him to beatbox *and* sing simultaneously.

---

[25] Fat Boys were a hip-hop trio active in New York City during the mid-1980s; known both for the aforementioned forays into beatboxing and the 'novelty' value of cuisine-centered hits (e.g., *All You Can Eat*).

[26] Stowell and Plumbley, *Characteristics of the Beatboxing Vocal Style*.

[27] Ibid.

[28] Thompson, "Beatboxing, Mashups, and Cyborg Identity."

[29] Rahzel Manely Brown (1964–) is an American beatboxer and rapper, was a member of 'The Roots,' and is known for his ability to rap and sing whilst beatboxing at the same time (e.g., *Iron Man*).

The astonishment of newcomers to beatboxing culture can be partially understood within the context of Denis Smalley's[30] proposed dichotomy between the 'gesture' and 'texture' of sounds: Richard Causton summarizes this theory by stating that "*texture* describes the internal life of sound as a static object, whilst *gesture* is concerned with its 'energy profile' or dynamic patterning." By this reckoning, beatboxing should be thought to have a gestural similarity or energy profile similar to that of the mimicked drum machine sounds. The same could be said of the outlandish 'DIY human synthesizer' known as Dokaka,[31] a Japanese 'viral' sensation whose multi-tracked, vocal-only covers of rock titans like Led Zeppelin and Metallica uncannily harnessed the acoustic energy of all the instruments being vocally mimicked while, again, not exactly being a believable stand-in for the real thing. Having said all this, though, a more simple explanation for this technique's enduring popularity might be its illustration of pseudo-polyphony, that is, the phantom presence of multiple instrumental voices where only one is present (and, again, it is wise here to conceive of the drum machine being mimicked as a synthesized collection of percussion instruments rather than a single voice).

5.4 *CyberSongs*—Text-to-Speech-to-Song Pieces

My fascination with the inventiveness and playfulness of vocal forms such as *katajjaq* or that of the art of beatboxing, in its all-encompassing use of all kinds of interlocking sounds and noises to generate sonic phantoms, eventually led me beyond the use of human voice to experimentation with computer 'text-to-speech' sampled voices as tools to explore different phenomena through the intensive and extensive use of repetition. One of these is the phenomenon of semanticization,[32] that is, giving meaning to arbitrary meaningless sounds after prolonged repetition. A closely related effect is that of musicalization or the 'speech-to-song' effect: generating rhythmic structure and apparent melodic lines and harmonies of non-musical sound material. Interestingly, we can also generate the opposite psychological phenomenon, known as 'semantic satiation,'[33] which implies a dissipation of meaning, as a consequence of the repetition of meaningful structured sound, particularly words. This is similar to the 'mental fatiguing' effect of continually staring at a word or phrase until its familiar meaning evaporates; over time a listener will perceive a repetitive spoken word stimulus as a meaningless sound or series of sounds.

---

[30] Denis Smalley (1946–) is a New Zealand-born composer and musicologist; after studying under Oliver Messiaen he made significant contributions to the concept of 'spectral music' or 'shaped' sound.
[31] Dokaka, known as a DIY human synthesizer, is also known as the Japanese beatboxer whose vocal-only reinterpretations of Led Zeppelin, Slayer, the Rolling Stones, and others gained cult status when his mp3s went viral. www.dokaka.com. Accessed October 20, 2019.
[32] Beheydt, "The Semantization of Vocabulary in Foreign Language Learning."
[33] Jakobovits, "Effects of Repeated Stimulation on Cognitive Aspects of Behavior."

Vocal Phantoms is a series of pieces that focuses squarely upon the usage and interpretation of patterns of speech for the generation of rhythm and meaning, as well as the antithetical breakdown of meaning, from the relentless repetition of morphemes, phonemes, words, sentences, and phrases. These pieces exploit this entire spectrum of linguistic and syntactical units as compositional elements: from micro-fractions of syllables to entire phrases, exploring the way in which the interplay between such speech patterns at different scales manipulates our attention and forces us to continually adapt our listening focus to accommodate fresh information. All these phenomena are, of course, reminders of the 'predictive brain' at work, and of the push and pull between top-down and bottom-up forms of sensory processing. These pieces play with rapid switching between different, occasionally competing, types of attentive listening. For example, at one level we might hear a recognizable word and on another level we might perceive phonemes, morphemes, or formants (i.e., the spectral characteristics of vocal sound); whilst on yet another level we might hear only acoustic features of a sound: its frequency, amplitude, duration, and so on. At each level, the sounds generate their different meanings (acoustic, semantic, informational), which are entirely dependent on the context or framework in which they unfold. As we will soon discuss with the phenomenon of semantic satiation, the meanings attached to words, objects, or events can degenerate at an astonishing speed when their surrounding framework and socio-cultural context disappear.

While a piece such as *Vocal Phantoms #18* is constructed from meaningless vocal units (in standard semantic terms), I have developed an entire ongoing sub-series of compositions called *CyberSongs* that explore the striking mutability of meaning, in both semantic and musical terms, that unfolds when working with vocal phantoms produced with the use of 'text-to-speech' computer technology combined with the 'speech-to-song' phenomenon.

It is a fascinating phenomenon that virtually any fragment of spoken speech, when looped and repeated, inevitably acquires a 'musical' character; words transforming into song, so to speak. The intensely looped cut sequences of seconds of film footage by Austrian experimental filmmaker Martin Arnold (e.g., his *Passage à l'acte* [1993]) are a powerful and clear example of this phenomenon. Diana Deutsch, well known for her research on musical illusions and paradoxes, demonstrated this effect in a series of experiments on what she called 'the speech-to-song illusion'.[34] As anybody with experience in the work with looped audio samples knows very well, this process of 'musicalization' is not exclusive to recorded speech. It is, however, particularly marked, apparent, widespread, and even probably consubstantial, with this type of material. Any particular sequence of speech sounds can appear to be heard as either speech or music (in that sense of being 'musicalized'), depending on the amount of repetition of the word or phrase considered. If the spoken text is played once, it is naturally heard

---

[34] Deutsch, Shermer, and Skeptics Society, "Phantom Words, Auditory Illusions, and Other Curiosities"; Deutsch, "Phantom Words and Other Curiosities"; Deutsch, *Musical Illusions and Paradoxes*; Deutsch, *Psychology of Music*.

as spoken text, but after being repeated a number of times it will be heard as music, complete with its own melody and rhythm. Since we are used to hearing repetition in music but not in everyday speech, the speech-to-song phenomenon might in fact reveal that there somehow exists a latent or background musical way of listening. We thus shift from hearing the words as words to hearing them as if they have marvelously and playfully transformed into melody and rhythm. Once this musicalization transformation takes place, the perceptive focus switches to the musical-sounding nature of the material.

In the *CyberSongs* series of compositions I have extensively and obsessively played with—among other things—this process of musicalization, through the use of standard text-to-speech computer utilities. Text-to-speech (TTS) is a type of speech synthesis application that is used to create a spoken sound version of raw text in a computer document. TTS (now used widely in the gaming and animation industry) is essentially a semi-artificial production of human speech from converted typed text and was originally implemented as assistive technology to enable, for example, the reading of computer display information for the visually impaired, or to generally aid in the reading of a text message. Samples of recorded chunks of speech or entire words and sentences are stored in a database, and then linked together to recreate the

Figure 5.3 Live performance with real-time video projection of *CyberSongs* (part of the compositional series *Vocal Phantoms*—Text-to-Speech-to-Song) at Störung #9 Festival, Barcelona, Spain, 2014. For some sections of this composition I created several female visual avatars to 'perform' the fast-paced, highly looped voices of the piece. Photograph by M. A. Ruiz © 2014 Störung.

text selected. I have mainly used TTS applications with sample-based real voices, as opposed to synthesized voices, to compose my pieces.

Unlike repetitions with manual or instrumental means (as in my *Harp Phantoms* and *Drawing Phantoms* compositional series), the use of text-to-speech utilities, combined with digital audio editing techniques, makes it possible to create sample-accurate copies of speech sound patterns, allowing an exact repeatability of any audio text fragment. A digital clip can be created and copied as many times as required or desired for the piece. In these ideal conditions for extreme repetition, despite the fact that all copies of the repeated fragment are identical, the perceptual experience of the listener is never one of static uniformity, but rather one of continually shifting changes. As mentioned before, it is precisely the use of repetition that engenders this transformation of a shifting perceptual experience. It is us as listeners, therefore, who do not remain the same, and each time we hear the repeated segment again a combination of expectation and memory reframes and transforms our experience of what we hear. Our normal speech patterns consist of sequences of known signals and we have come to expect certain sequences of patterns and arrangements of signals (words, sentences) according to our given rules of speech. When we are exposed to repeated identical signals our brain eventually finds alternatives, as it tends to expect the next sequences to be different to what took place immediately before.

Getting the computer to read intricate (both meaningful and meaningless) textual patterns with a TTS application and recording and editing the audio output, I have created an extensive repository of sonic micro-loops (in the range of 50 to 500 milliseconds) for the purposes of making these text-to-speech pieces, with various combinations of patterns and layers of these small audio units in synchronous and asynchronous organizations. The *CyberSongs* pieces have been composed in this manner, using these techniques as a way to increase the complexity of the initial material and to explore forms of considerably intricate polyphony that are possible only with digital tools.

Although on some occasions I use complete words, most of these micro-TTS loop units and the small repetitive patterns created with them are meaningless letters, phonemes, or sequences of both that were chosen or produced essentially on the basis of their timbral or percussive qualities. Each composed pattern is continuously repeated in a regular fashion and forms a distinct layer with its own morphological characteristics. Some of the simpler constructions constitute what we could call 'monomania' pieces in their own right; simple but full of phantom-inducing power. In these more stripped-down compositions an initial sequence is composed and built up to form a stable pattern. This pattern is then slowly transformed in some manner; for example, deconstructing and contracting the loop as new samples are gradually introduced to replace existing ones.

In the more complex and more densely layered pieces, I have combined such layered patterns in different configurations, stacking and offsetting them against each other to create densely detailed textures. The symmetry and alignment of layers in their simultaneous organization, as well as their composed internal microstructure, have a paramount effect on the complexity of the emergent sonic image. As in the case of the looping process used in the *Harp Phantoms*' compositions, the construction of these pieces involves first setting up at least two identical (or almost identical) patterns to initially play together in synchrony, aligned and panned to center position. The purpose of this is to first establish a 'ground' pattern that is presented over a sufficient duration of time to become stable (this duration varies depending on the sonic features of the pattern itself). Once this ground pattern is stable enough to be internalized in the short-term listening memory, a replica of it is added in a different layer, but with a micro-shift in time with respect to the first one. Layers can be shifted or nudged incrementally to very fine degrees of resolution using digital software tools. For reasons of aesthetics and of perceived efficiency in generating sonic phantoms, my usual choice of average time shift between layers in these pieces has been in the range of 10 to 50 ms.

In summary, by combining all the technological, perceptual and compositional strategies described above, the *CyberSongs* suite generates a creative techno-cognitive-musical link that stretches through a compelling, syncretic, and seamless chain that we could call 'text-to-speech-to-song.'

5.5 Micro-Temporal Mechanisms

The techniques described above for my *Vocal Phantoms* pieces, as well as those employed in traditional *katajjaq*, can be contextualized and better understood in terms of perceptual mechanisms in the light of the research and experiments carried out by Diana Deutsch on what she called 'phantom words illusion.'[35] In these experiments, Deutsch used looped recordings of two words or a single word with two syllables, presented separately but simultaneously through a stereo pair of loudspeakers. The two sequences of sounds are played simultaneously on both speakers but the tracks are offset in time so that when the sound of the first syllable is playing on the left speaker, the sound of the second syllable is simultaneously playing on the right speaker. She explains that

> because the sounds coming from the two loudspeakers are mixed in the air before they reach our ears, we are given a palette of sounds from which to choose, and so can create in our minds many different combinations of sounds.[36]

Since the same words are repeated and displaced between the two channels, this causes ambiguity and confusion and so we can hear all kinds of shifting words and phrases that are not really there, as our cognitive system attempts to make sense of the aural ambiguity. These experiments help to clarify how the sonic phantoms emerge

---

[35] Deutsch, *Musical Illusions and Paradoxes*; Deutsch, "Phantom Words and Other Curiosities."
[36] Deutsch, "Phantom Words | Psychology Today."

from the *katajjaq* or my own *Vocal Phantoms* pieces. One has the impression that the two voices are producing two different series of sounds, when in fact both voices are creating each of the sounds heard. Both the spatial distance and the timbral similarity between the voices are critical for the illusion to work.

In constructing my *Vocal Phantoms* pieces I regularly use this form of structural organization of two or more sequences of looping sounds; that is, using two identical copies (or a slight variant) of the same loop and presenting them with offset to both channels left and right simultaneously. These sequences of sounds therefore arise simultaneously from different regions of the space. In general, as each loop motif contains a low-pitched sound followed by a high-pitched one (or the contrary), and since the two tracks (or live voices) follow each other with a time lag, we have the impression of hearing two separate distinct streams, one composed of low-pitched sounds and a second one of high-pitched sounds, even though both types of sounds have been produced alternately by each voice track (or live voices). The aim is that the resulting emergent sound image produces such homogeneity of sound that listeners or spectators are not able to discern the separate sources.

There are some important 'ground rules' regarding looping techniques that should be laid out here. As already mentioned in chapters 3 and 4, for sonic phantoms to emerge there must be a sufficient number of repetitions of the looping material: there is a threshold at which the number of loop repetitions, combined with the number of elements in the loop itself (the motif), will be sufficient for the induction of phantom patterns. Below this threshold, there will not be enough time for the perception to reach that 'phantasmatic state.' With sufficient exposure, as the primary motif patterns are internalized, perception is freed from having to concentrate on those particular patterns and it has the free cognitive space to invent and create new ones. These new secondary patterns are automatically picked out and brought to attention via the shifting of figure-and-ground relationships amongst the patterns. Each new pattern that is foregrounded into conscious awareness and then internalized brings about more dissociation, and all these multiple dissociations provide more chances for the hearing of sonic phantoms.

Again from trial and error, and while experimenting with different lengths by judging results with the 'perceptive' ear (i.e., in terms of the sonic illusions), I have observed that the combination of the length of the loop itself and the number of differentiated sonic elements it contains are both critical to attain the effect of its dissipation into independent streams of sound. When there are too many elements in the loop, or their independent durations are too long, this will weaken the effect of dissociation. In these kinds of pieces, too many elements in the loop make it difficult for the brain to internalize the patterns or retain them in memory. Shorter looping structures are in general more effective, and any attempts or strategies to achieve mechanical precision, even when performed by live voices, will greatly contribute to the effect.

In terms of the average duration of the loop units used in my *Vocal Phantoms* pieces (i.e., in the range of 50 to 500 ms, as mentioned before), this work could be located within the temporal realm of granular synthesis' 'microsound,' also described by composer Brigitte Robindoré[37] as 'quantum sonics' (wherein the average sonic 'grains' are in the range of 10 to 100 ms: the title of a piece by microsound champion Curtis Roads,[38] *Sonal Atoms*, provides an adequate allegory for this means of quantifying sonic units).[39]

All concerns relating to temporality aside, the content of any loop motif needs also to be carefully considered, so as to be eventually repeatable by the mind, and thus provide strong sonic phantom-inducing structures. As discussed previously, the semantic content of the loop itself does not influence the effectiveness of the trance-inducing process. Any arbitrary or random choice of meaningless content, therefore, will in principle serve the purpose of inducing a state of deep absorption (at least at a primary level). From a sonic or aesthetic perspective, however, the content of the loop will naturally affect the timbral and rhythmic quality of the phantom patterns. This will then affect how interesting and engaging the experience will be for the listener. Consequently, although the purely sonic content of the loop will not affect the initial induction, I have experimented with gradually transforming the looping content elements over time to offer a sonically richer aesthetic experience.

A similar aesthetic to what has been mentioned here, though originally materialized using the versatility of the human voice rather than the chattering digital textures of granular synthesis, comes from the work of the trailblazing French sound poets Henri Chopin and François Dufrêne,[40] who took the written phonetic poetry of the Dada *provocateurs* (Hugo Ball, Raoul Hausmann, Tristan Tzara, etc.) and the theories of the later Lettrist movement more solidly into the sonic realm. Chopin in particular did this both by crafting poems that acted as spiritual successors to this work, and by being one of the few to record or publicly perform the works of Hausmann and Kurt Schwitters.[41] The larger movement that Chopin and Dufrêne represented also perceived individual sonic units—be they vocally produced or otherwise—as 'objects' to be distributed throughout space much in the same way

---

[37] Brigitte Robindoré (1962–) is an electroacoustic composer and student of GRM (Groupe de Recherches Musicales), studying under luminaries including Iannis Xenakis and Jean-Claude Risset.

[38] Curtis Roads (1951–) has made some of the most significant contributions to computer-based music in the late-twentieth and early twenty-first centuries, both with his own forays into granular synthesis and his editorial role at *Computer Music Journal*.

[39] The piece in question appears on the 2001 compilation album *New ElectroAcoustic Music from Paris*, released by CCMIX (Center for the Composition of Music Iannis Xenakis in English and Centre de Création Musicale Iannis Xenakis in French).

[40] François Dufrêne (1930–1982), associated with French avant-garde thinktank the Lettriste International, pioneered a form of non-verbal sound poetry known as 'crirhythmes.'

[41] Kurt Schwitters (1887–1948) was a German artist active in numerous fields, and is widely known for his *Merz* collages of found materials and for *Ursonate*, the 'primeval' vocal sonata.

that practitioners of *musique concrète* had done[42] (the shift arguably dates to Öyvind Fahlström's[43] 1963 radio play *Fäglar i Sverige* ['Birds in Sweden']).

The works of Chopin and Dufrêne have been described by Charles Amirkhanian as "sound poems which deal in sonic units smaller than syllables."[44] Chopin has, in turn, referred to his utterances as 'micro-particles,' basic units charged with an elasticity of interpretation that prevents them from being distinctly classified as poetry or music. The comparison to computerized microsound may, initially, sound like a bit of a stretch, yet a choice Chopin composition such as *Lé Ventre de Bertini* ('Bertini's Stomach'), using little more than a superimposed succession of glottal vocal utterances, quite elegantly approximates the characteristics of that much later psychoacoustic innovation (while also approximating a particularly turbulent bout of intestinal activity). Other examples such as *Les Mandibules du Déjeuner sur l'Herbe* ('Mandibles of Luncheon on the Grass') rely upon a similar technique, compiling dozens of individual micro-phonetic utterances into a mass of 'insectoid' sound that mischievously plays upon the Manet painting referenced in the title: the sound poet's machinations make listeners understand that the 'luncheon on the grass' can refer not exclusively to the painted human subjects who are resting upon the grass while they eat, but to other life forms making that same grass into the main course for a luncheon of their own.

Micro-phonetic extended vocal technique (or perhaps 'pure phonism,' as Raoul Hausmann would have called it) is not at all limited to the artists already mentioned here; in fact, the task of forging inspirational works out of this reductionist tendency has been shared throughout the history of sound poetry. Larry Wendt,[45] for example, was intrigued that "a word which had a particular semantic sense while whole could produce a variety of sonic environments when fragmented."[46] Wendt's own experiments have occasionally gone in this direction, while also playing with the possibilities presented by computer processors' characteristic misinterpretation of recognized phonetic structures (the apparent fragility of technology, and the absurdist consequences thereof,

---

[42] This comes with the caveat that there was a significant amount of disagreement between *musique concrète* founder Pierre Schaeffer and the representatives of text-sound or sound poetry. As Larry Wendt recalls, "Schaeffer believed that sounds, as concrete sonic events, could be pried loose from the physical acts that created them and given a new significance as a result of their placement in an abstract musical composition. The works of these experimental poets, on the other hand, emphasized the fact that vocal sounds are created by the body, and that an act of creation provides the structure for the arrangement of these sounds in a composition. Schaeffer has largely recanted his earlier theories about abstracting sounds for compositions and assumes the stance of some of his critics (such as French Structuralist Claude Levi-Strauss)." Wendt, "Vocal Neighborhoods."

[43] Öyvind Fahlström (1928–1976) was a Swedish artist and playwright specializing in sociopolitical content.

[44] Ode to Gravity: Three French Sound Poets, 1972, https://archive.org/details/OTG_1972_09_27_c1. Accessed December 1, 2018.

[45] Larry Wendt (1946–) is a text-sound artist and writer; active since the mid-1970s, his work has included numerous incisive pieces dealing with the realities of life and work in Silicon Valley.

[46] Wendt, "Sound Poetry: I."

are recurring motifs in Wendt's work that stem from his experiences within Silicon Valley as it existed during the dawn of the microcomputer market). This is the case with some of the pieces compiled on his 1979 cassette *Sound Poems for an Era of Reduced Expectations*. A separate experiment of Wendt's, from several years later, uses multiple layers of semantic breakdown to create odd hybrid phonemes:

> Another device I have constructed and used is a text-to-speech synthesizer which allows one to use a keyboard to type texts which the synthesizer attempts to say. One such study of mine using this device was *Starting from Maya* (1984). The text for this piece was an excerpt of transliterated Mayan writing. I typed the text into the text-to-speech synthesizer and it pronounced it with English phonemes to generate nonsense words. I then rewrote the texts several times, gradually replacing the nonsense words with the real words and phrases which the sounds reminded me of. I made two variations on the texts and the two resulting synthesized voices were super-imposed upon one another in a sort of question-answer catechism.[47]

This experimental urge is described further by Bernard Heidsieck[48] while in conversation with Amirkhanian. Here, a kind of revelatory quality is being ascribed to the process of de-semantization that follows in turn from a micro-temporal approach:

> HEIDSIECK: If you read something and, for instance, a word—if you just give half of the word, the people listening to that put the end of the word immediately into their mind without hearing it. And so, you can make a piece by …
> AMIRKHANIAN: Yes, I understand. It also gives a very interesting quality to the sound to hear half the word. So you have two benefits—one, you move people along very quickly, but you also give them a sound which is very unusual. To hear half a word—nobody ever says half a word, like "ha-"![49]

So, after a fashion, there is some 'expansion through reduction' going on in the semantic alterations that Heidsieck proposes. This manages, on a purposive level, to unite the 'microsound' of the French sound poets with that produced by the granular synthesis engines and algorithmic processes employed by Roads and like-minded composers, in that the conceptualization of sonic units as being much temporally smaller than usual makes it paradoxically easier to portray a world in flux or a state of streamlined continuity. As Roads suggests, "the flowing structures that we can create with microsound do not necessarily resemble the usual angular forms of musical architecture … to the contrary, they tend toward liquid-like or cloudlike structures."[50]

---

[47] Wendt, "Sound Poetry: I."
[48] Bernard Heidsieck (1929–2014) was one of the original sound poets, creating personalized categories for his work that included 'biopsies' and 'poem partitions.'
[49] Ode to Gravity: Three French Sound Poets, 1972, https://archive.org/details/OTG_1972_09_27_c1. Accessed December 1, 2018.
[50] Robindoré and Roads, "Forays into Uncharted Territories."

This theme has gained serious traction over the past few decades, affirmed by Trevor Wishart's suggestion that "we can... create imaginary phonemic objects and in fact construct imaginary languages and linguistic streams,"[51] with new compositions regularly appearing that make good on this promise.

5.6 Semantic Satiation and Semantization

The conventional wisdom regarding monotony is that "most organisms stop responding to a stimulus repeated over and over again (unless the response is reinforced by reward or avoidance of punishment)."[52] This is apparently borne out by the realization that "higher organisms actively avoid a completely monotonous environment"; for example, rats in a maze will take alternating routes to achieve the same goal of attaining food, though knowledge of a single route will be perfectly sufficient to realize this goal. This begs the question of exactly what happens when we are compelled to experience a monotonous environment and do not have the kind of option accorded to our lab rats above. Numerous experiments in sensory deprivation, from those carried out at McGill University in the 1950s to the later 'Ganzfeld' experiments,[53] suggest a plethora of intense reactions to environments that are effectively devoid of any perceptible sequence of differentiated events. From reaching a state of complete cognitive emptiness (i.e., being unable to think of anything at all), to hallucinatory experiences rivaling those normally associated with potent plant hallucinogens, most of these experiments have in common the conclusion that a 'pure' state of phenomenological monotony is difficult to maintain. As we will see, this relates directly to audio content whose variety is severely limited: the 'audio-phantasmatic' finds unexpected ways to bloom even when semantic content is strictly limited (and, as we will see, often precisely *because* it is strictly limited).

In Edgar Allan Poe's short horror story 'Berenice' (first published in 1835), Egaeus, the protagonist, who has a tendency to fall into periods of intense focus often leading to dissociation, describes a psychological state, a 'monomania,' that amongst other things induced him "to repeat, monotonously, some common word, until the sound, by dint of frequent repetition, ceased to convey any idea whatever to the mind."[54] While presented in the gothic horror of Poe as a morbid act or pathological condition, we have already seen from discussion of mantras and similar poetic practices that such monomania can also be a voluntary act that aims at the *opposite* of the obsessiveness

---

[51] Wishart, *On Sonic Art*.
[52] Heron, "The Pathology of Boredom."
[53] This experiment can be carried out by, simply enough, placing the two halves of a dissected ping-pong ball over test subjects' eyes (generally these will be somewhat sculpted to conform to the dimensions of the eye sockets). Though it has been introduced as a means of examining neurophysiology in general, researchers in the field of parapsychology have had their own uses for it. See Carpenter, "ESP Findings Send Controversial Message."
[54] Poe, Levine, and Levine, *The Short Fiction of Edgar Allan Poe*, 72.

suggested by such fictions. The realm of modern conceptual art is rife with practices that involve flipping obsession on its head and, in the process, revealing an aesthetic or self-critical dimension in which monotony is taken to its perceptual limits, or in which the aforementioned triumvirate of dissociation, disorientation, and fragmentation become analytical tools rather than destructive means of 'breaking down' consciousness. The Beijing-based artist Qiu Zhijie,[55] for example, has demonstrated this in his calligraphic work *Assignment No. 1: Copying the "Orchid Pavilion Preface" 1,000 Times*, a video-recorded piece that involved endlessly repeating the most classic calligraphic text of Chinese tradition on a single sheet of rice paper. *Assignment No. 1* was described by the artist as follows:

> As the text is written the first time, the calligraphy is clearly of Chinese characters. As the number of times increases, the characters are turned into a visual merging of ink, like an abstract painting. Following the first fifty times during which the characters become obscured, the text is rewritten on the paper, which gradually turns black with no trace of the brush mark or nuances of ink… *calligraphy is made Zen* [italics ours].[56]

The same effect often occurs within the realms of 'language poetry' or the 'conceptual writing' project championed by Marjorie Perloff[57] and Kenneth Goldsmith,[58] whose radical claims (i.e., "one doesn't need to generate new material to be a poet" and "the intelligent reframing of extant text is often enough")[59] regularly manifest themselves in dizzyingly repetitive works where the reader or listener is barraged with slight semantic variations on a theme. Through pieces such as Alexandra Nemerov's 2007 poem *My First Motorola*, the instrumental qualities of language (poetic or otherwise) gradually dissolve and the constituent items within exhaustive lists of objects, events, or impressions flow together into a singularity often devoid of message (in the case of Nemerov, the effect is achieved with a litany of brand-name consumer goods). This approach is utilized in the culture of sonic art, as well, with many of the examples in the 'persistence' section of this book qualifying. Some works, spanning entire album lengths, follow the spirit of semantic satiation, if not quite the letter: the 2007 CD release *An Archivist's Nightmare* by experimental musician RLW (a.k.a. Ralf Wehoswky)[60] offers one striking manifestation of this, consisting as it does of the artist

---

[55] Qiu Zhijie (1969–) is a Chinese media artist and educator concerned with issues of the information age, e.g., semiotics and ephemerality.
[56] Hung et al., "Reinterpretation."
[57] Marjorie Perloff (born Gabriele Mintz, 1931–) is a poetry critic notable for proposing connections between the poetic avant-garde and postmodern literary theory.
[58] Kenneth Goldsmith (1961–) is an American conceptual writer and poet, as well as founder of the UbuWeb avant-garde archive online and the PennSound poetry archive.
[59] Millar, "Conceptual Writing."
[60] Ralf Wehowsky (1959–) is a German sound artist previously active in the post-industrial group P16. D4 and the audiovisual group Selektion.

flatly intoning the latest additions to his sizable avant-garde record collection, without additional qualifying commentary, for a full hour.

Though here we may be getting ahead of ourselves, since the phenomenon known as semantic satiation—as originally proposed by Jakobovits in 1962[61]—does not require constructions nearly as complex as the above to work its effects, and can in fact occur with no more than repetition of a single word. It is quite remarkable that to increase familiarity with a word during a learning process repetition is required, but with too much repetition a threshold is eventually reached where this cognitive pattern becomes suddenly defamiliarized. Within the same type of repetitive process, what was once a regular ordinary word with attributed meaning is suddenly transformed into a kind of strange phantasm of sound.

So-called 'minimal techno' music (which should be differentiated from the more melodically-driven 'EDM,' 'Electronic Dance Music,' confused for 'techno' by casual listeners but disdained by more hardcore adherents of the form) again provides an example of how this has seeped into pop culture. Countless techno tracks feature indiscernible and disembodied vocal samples, rarely exceeding five spoken syllables in length, which are welded to the pitiless and unerring 4/4 rhythmic structure of the music. After enough insistent repetition, these spoken snippets tend to progress from possessing a clear meaning, followed by a period of doubt in which a string of alternate interpretations arise, and eventually collapsing on themselves until they are nothing but another purely musical element wherein qualities of melody, timbre, and so on are to be contemplated instead of linguistic content. This perceptual transformation makes them, effectively, into miniaturized 'sonic blotscapes' within a larger musical framework.

Both the 'monomania' that Poe describes in his story, and the sampled techno chatter noted above, are contrasting examples of 'semantic satiation,' a psychological phenomenon in which the uninterrupted repetition causes a word or phrase to temporarily lose meaning for the listener.[62] This phenomenon highlights the arbitrary nature of our conventions to connect and associate certain sounds with certain meanings: anyone familiar with the experience of repeating a word over and over again, until the word becomes strange and meaningless, can attest to this. Even words loaded with emotional content, like 'mother' or 'blood,' words that are normally for us solidly familiar and clear, uncannily transform themselves in our perception in strange phantasmatic ways. Virtually any word we choose will start to sound peculiar—in this particular way of dissipation of meaning—if we repeat it audibly enough times. This phenomenon creates an in-between state of unfamiliarity and strangeness; one that hovers in a temporary zone between meaning and sound. The act of writing a word or a phrase repeatedly has a similar effect; one of dissociation, disorientation, and fragmentation.

[61] Jakobovits, "Effects of Repeated Stimulation on Cognitive Aspects of Behavior."
[62] Professor of Psychology Richard M. Warren at the University of Wisconsin-Milwaukee conducted perhaps the most in-depth scholarly research into this phenomenon also in the 1960s, which he referred to as 'verbal transformation effect.' See Warren, "Illusory Changes of Distinct Speech upon Repetition."

Psychological researcher Sheila Black reaffirms that excessive exposure to a single word is all that is needed to demonstrate a concurrent decrease in its meaningfulness. In the process, she also describes the evolutionary usefulness of the phenomenon, that is, its existence as a mechanism "to dampen the potency of redundant information, thereby reducing the amount of clutter in working memory from stimuli that have little information value."[63] Following from this, Black also notes how the passive or unconscious nature of semantic satiation provides it with a unique character relative to other forms of inhibitory defense. She also, via discussion of semantic activation theory, provides a more granular understanding of the mechanisms that give this effect its potency, suggesting that mental structures called 'nodes' are responsible for the feeling of fatigue arising not only from distinct words like 'doctor', but also from words that may be associated with the already identified fatigue-inducing word (e.g., 'nurse').

As the level of excessive exposure is ratcheted up, these nodes—which have also been referred to as 'm-components' by researchers Kanungo and Lambert—will gradually 'drop out'; that is to say, the original word will seemingly devour other potential associations (in their example, "repeating 'table, table, table' will increase the tendency for 'table' to evoke 'table' as an associative response").[64] This sort of effect seems to be illustrated by Paul Sharits's structural film *T,O,U,C,H,I,N,G*, in which a couple of minutes' worth of the repeated verbal command 'destroy' jockeys for sensory attention with the image of a single male figure framed against a series of strobe-lit, bright background colors: it is somewhat unsettling how the insistence of the audible command effectively manages to divert attention away from what might otherwise be a *creative* process (such as creating a sort of back story about this figure and the minimalist environment surrounding him) towards associations with *destruction*. Stranger still are the reports of viewers hearing other utterances—namely, 'the girl is gone' and 'history.'[65] These reported utterances seem to hint at the drop-out phenomenon just mentioned, in that these percepts could also indicate a negative state of loss arising from the previous concentration upon destruction.

There are a few active sound artists who refer explicitly—as is not the case for all of the examples above—to semantic satiation as a technique to be used in their own work or explored by others. This is the case for sound artist and theorist Michel Gendreau, who discusses the compositional uses of semantic satiation in his book of aphorisms, *Parataxes*. Suggesting that it is "the opposite of temporal persistence," he also notes that its phantasmatic effect is strong enough to achieve what was mentioned earlier in regards to auditory driving—that is, semantic satiation provides listeners with "the opportunity to rethink a stimulus"—and then goes into further detail about how it can be deployed compositionally:

---

[63] Black, "Semantic Satiation and Lexical Ambiguity Resolution."
[64] Kanungo and Lambert, "Semantic Satiation and Meaningfulness."
[65] Smith, *Muses, Madmen and Prophets*.

Semantic satiation, in an odd reversal of the active role of the composer, provides variations through the varying contextual environments and the changing perception of the listener (the composer controls other internal changes *behind* the loop only). The repetitive element becomes familiar, and thus more or less transparent, making it easier to resolve the results of its combination with the perceptual or environmental elements. Not just any loop can work in this way, because the combination of arbitrary elements does not systematically produce a useful result, although the essential interaction with human perception is somewhat forgiving (for better or for worse).[66]

To better describe the potential perceptual effects of this phenomenon, Gendreau also refers to Alexander Potebnya's claim that "a thought once connected with a word is again called into our mind by the sounds of the word … the thought is reproduced not in its previous form but so that the second and third reproduction may be even more important to us than the first one."[67] Following this line of thinking, it is easier for us to understand how solitary atoms of insistently repeated verbal information could, having reached an apex of being meaningful, disintegrate so that perceptual focus shifts to their sonic properties rather than their semantic (linguistic) content. Gendreau's later assertion that "thinking is continuous, but language is incomplete" also, in its own way, describes this capacity for perception to eventually outpace meaning.

In a less scientific and more purely aesthetic sense, semantic satiation could be seen as a recent iteration of the Russian Formalist Viktor Shklovksy's[68] aesthetic theory of *ostranenie* ('making strange'), as well as being a proof of the same. In Shklovsky's reckoning, intense concentration upon the customary and the near-at-hand 'defamiliarizes' it in such a way that its original construction can be more fully apprehended. To wit:

> The purpose of art is to impart the sensation of things as they are perceived and not as they are known. The technique of art is to make objects "unfamiliar," to make forms difficult, to increase the difficulty and length of perception because the process of perception is an aesthetic end in itself and must be prolonged.[69]

It is certainly beyond the scope of this book to theorize the possible reasons—such as the critique of the information age's relentless anti-essentialism—for an increased interest in semantic satiation as an incisive creative tool. Suffice it to say that audio artists have been making use of it, and even calling it by its name, in times when the infosphere was far less saturated than it is at present, one notable example being

---

[66] Gendreau, *Parataxes*.
[67] Ibid.
[68] Viktor Shklovksy (1893–1984) was a Russian writer and theorist active in the Formalist movement, which had considerable influence upon later structuralist and post-structuralist theories, from the 1910s to 1930s.
[69] Shklovskii et al., "Russian Formalist Criticism."

composer Nicolas Collins with his 1986 piece *Vaya con Dios* ('God be with you').[70] Other pieces from Collins's back-catalog are also representative of an aesthetic of satiation/saturation, such as the compilation track *Devil's Music 1*: a sonic barrage of turntable scratches and speech granules in which "the needle-like precision samples are so short as to erase any meaning the sound excerpts might have for the listener... you're left with pure meaningless aural stimulus... arguably what the composer think[s] of mass media Muzak."[71] The presence of semantic satiation effects is also implied by the theories of electro-acoustic musician and educator Denis Smalley, particularly his emphasis on the use of Schaefferian 'reduced listening'[72] as a kind of pedagogical tool by which listeners can train themselves over time to disassociate sounds from their original sources.

Interestingly, when semantic satiation, normally confined to language, is brought into an aesthetic or musical realm, the dissipation of (semantic) meaning can give rise subsequently—and somehow paradoxically—to a new meaning that is relative to this new cultural frame of reference. This is what happens with the eventual musicalization of those intensely repeated language elements that have already lost their original, usual meanings. Processes of semantization, therefore, not only take place when 'abstract' sounds turn into 'words'—as in EVP and in the *Drawing Phantoms* pieces (OVP) described above—but also when sounds that have been 'abstracted' from language turn into 'music.'

In the *Vocal Phantoms* pieces I use hyper-repetition of speech sonic patterns in various combinations to, among other things, perceptually shift our attention to 'semantically hidden' timbral aspects of the sounds; sonic perceptual layers of the sounds that under normal (linguistic) listening conditions we would not be able to hear. In this way, we switch to listening modes that actively attend to the timbral and sensory qualities of the sound; sonic textures, tones, melodies, rhythms that we never heard before. Music that we had not heard before, coming from 'inside' those words.

A particularly relevant version of this musicalization takes place in some of my *CyberSongs* pieces, in which I used computer text-to-speech computer voices from different nationalities. The available options for the choice of voice in text-to-speech software (with personalized names such as 'Alva,' 'Kyoko,' or 'Esther') are obviously provided to render a naturalized voice result when using the right match of language between written text and spoken voice. When seen through a musical prism and approached with a playful attitude, it was just natural for me to take a step beyond the intended use of these tools, and to experiment with the tempting mismatches

---

[70] Collins claims that, in this piece, "a digital delay loop is used to achieve 'semantic satiation'; any phrase repeated enough starts to sound meaningless." See Collins and Plsek, "Nicholas Collins Responds to Reviewer Tom Plsek."

[71] Ubuweb: Sound, Randy Greif, http://www.ubu.com/sound/tellus_20.html. Accessed December 20, 2018.

[72] In Schaefferian theory, reduced listening is the perspective of listening to the sound for its own sake, as a sound object, by removing its real or supposed source and the meaning it may convey. Schaeffer, *Treatise on Musical Objects*. See also Kane, *Sound Unseen*.

Figure 5.4 A different visual avatar I created to 'perform' (via Text-to-Speech-to-Song) a sound installation version with video projection of some of my *Vocal Phantoms* at the Modern Body Festival, The Hague, The Netherlands, 2014. Photograph by Thijs Geritz, 2014.

between both: I could get Alva from Sweden to speak Hungarian words; Kyoko from Japan tried her best to make sense of English; Esther from Hungary seemed to make some sort of sense out of any language. Besides the obvious plethora of potentially comical results, what these chimerical experiments reveal, and what the resulting pieces abound with, is the immense musicality of language prosody beyond what is normally perceived. I have freely used all imaginable combinations of languages, voices, meaningful text, and nonsensical fragments to generate very specific types of musicality, always aimed at inducing sonic phantoms. In many cases, the illusions are the result of apparent meaning, either temporary or persistent, when the listener thinks he/she can recognize certain words—or is trying to do so—and is therefore very actively and acutely exploring the voice patterns and all their sonic intricacies in search of possible meaning.

This particular acute state of listening, combined with the effects of repetition, was investigated by the scientist John C. Lilly in a famous experiment involving what he called 'the repeating word effect.' In this experiment he made an extensive study exposing over 300 subjects to a recording of the repeated word 'cogitate' (as a sort of tribute for this, I have created a piece that represents my own version of this experiment). Very much in keeping with the examples already mentioned, results showed that if participants listened to the word 'cogitate' for fifteen minutes or more (they tested durations up to six hours) then they heard on average twenty to thirty different words in addition to the one repeated in the recording. The listeners were requested to notate changes in sounds and meanings of what they heard. From the 300 experimental subjects they collectively notated 2,300 different words, 300 of which were dictionary words and the remainder of which were neologisms or nonsense words. Writer Edward Rosenfeld participated in the experiment himself and reported:

> "After a few minutes I became tense and nervous," recalls Ed, "I was tired of listening to the same word over and over. Suddenly the tape changed. It said, 'MELT INTO IT.'" After the tape was over, Ed turned to a friend to compare notes. "Tape changed, didn't it?" he said. "It sure did," said his friend, "It changed from saying 'COGITATE' to 'COUNT TO TEN.'" It turned out that everyone in the room heard the tape change to a different word or phrase, and all were convinced that the change was actually on the tape and not in their brains. However, the tape had never actually said anything other than "COGITATE." These auditory hallucinations are the brain's natural response to a boring, repetitive stimulus, according to Lilly.[73]

Coming back full circle to the message implied by the 'rats in maze' and their preference for continually switching up their routes, Lilly's explanation of the phenomenon points to how intolerant our minds are to repetition, and how they necessarily will produce an imaginary illusion of change. This could be plainly described as the brain 'getting bored' of hearing the same word over and over again; since it does not have to focus on new incoming material—as it normally occurs—attention shifts to a different level of listening. The continuous repetition of the word pattern will remove meaning and reference or symbolism, and so we move from hearing in a schema-based process to a more 'primitive' mode of listening.[74] We do not hear a word but rather a group of sounds, a number of auditory streams with different timbral qualities and rhythms. The word, eventually heard as nonsense, detaches itself from its semantic meaning due to dissociation and it is now reduced to its acoustic nature, where semantics vanish and are replaced by an exaggerated accentuation of the component parts, letters, phonemes, prominent syllabic chunks. This feeling of dissociation can last for a period of time, until context or framework returns the word to its former position.

In the context of 'auditory stream segregation,' the once coherent auditory stream of the familiar word begins to split into auditory sub-streams that force a focus on a different level of listening. The result is a focusing on the acoustic nature of the sounds, which split and fuse into a variety of distinct sonic streams. There are interesting implications that such a process has for the construction of meaning in general, particularly if we disregard the idea that each of our senses is an autonomous

---

[73] Rosenfeld, *The Book of Highs*, 246; Hooper and Teresi, *Would the Buddha Wear a Walkman?*
[74] Denham and Winkler, "The Role of Predictive Models in the Formation of Auditory Streams"; Handel, *Listening*; Bregman and Rudnicky, "Auditory Segregation."

realm and acknowledge their working in concert to give us a complete picture of reality. As the acoustic research community dives deeper into subjects so seemingly alien to us as infrasound and ultrasound —that is, forms of sound that can affect us physically without being consciously perceived—we would do well to understand just how alien the 'normal'—familiar, common, mundane, daily—forms of sound can already be.

Figure 6.1 Fieldwork location for environmental recordings and on-site listening at the Great Otway National Park, Australia. Photograph by Francisco López, 2013.

6

Phantasma Naturalis: The Realm of Nature
Compositional Series: *Natural Phantoms*

> *I am lying on the soft leaf-covered ground, deep inside the forest at night, listening profoundly in the dark for hours. Despite the archetypal combined forces of darkness and solitude, as well as very real potential hazards, this is not a frightening experience but a transcendental one. I hear hundreds of tree frogs all around in a three-dimensional, naturally immersive, invisible constellation of sounds. Complex, intricate, all-encompassing, mesmerizing. But is it really the frogs that I am hearing?*[1]
> —Francisco López

6.1 Listening and Recording

There is a certain line of thinking in the Western musicological tradition, dating back at least to seventeenth-century thinkers such as the Abbé du Bos,[2] in which "musical sound [is] no less than the voice of nature itself";[3] put another way, the former does not approximate the latter so much as amplify it or extend its reach and deepen the emotional affect of nature's sonic constructions. Likewise, as the above quote by López suggests, so much of our aesthetic experience is bound up not just in the act of approximating the sounds of nature, but in attempts to find a sort of symmetry between our own audial *Umwelt* and one that (in the case of López's frogs) has preceded our arrival by millions of years. Yet, more intriguing than this, and knowing what we know at this point of this study, is that the achievement of such a symmetrical condition is a gateway to phantasmatic effects.

Aldous Huxley, in arguing the essential differences between the human and animal life worlds, claimed that humans sought out a condition of 'grace' that was already inherent in animals. That is to say, the actions and communicative habits of the latter could be distinguished from the former by their naïveté; by an incapacity for matching the infinite variations on deceit (with self-deceit not the least among these) of which

---

Readers can listen to several pieces of this compositional series on the website www.sonicphantoms.com
[1] López, "Sonic Creatures."
[2] Jean-Baptiste Dubos (1670–1742), also referred to as l'Abbé Du Bos, was a French author.
[3] Mace, "Marin Mersenne on Language and Music."

humans were capable. This is open to debate, to be sure: the sonic character of the natural world is abundantly populated by examples that would contradict the spirit of this assumption, from mynah birds' uncanny ability to mimic the timbral quality of mechanical sounds to killer whales' penchant for sonically disguising themselves as barking sea lions (and such an inventory does not even take into account deceit by stealth; e.g., the ways in which the barbastelle bat modifies its echolocation to outwit the 'jamming' talents of moths). However, even if acceding to a kind of 'naïveté' as far as *intentional* means of deceit are concerned, it can still be argued that unintended forms of deceit within the natural world are manifold. That so many of these occur while sonically experiencing the natural world should come as no surprise. The pioneering ethologist Jakob von Uexküll, when introducing musical metaphors to describe the 'emergent' operations of nature (i.e., "animals create worlds as an unfolding not unlike the temporality of music"),[4] implied lifeworlds rich with actual music, and a music whose interpretation leaves the door wide open for intriguing phantasmatic sensations.

It is in this spirit that the realm of sonic experience that I have called *Phantasma Naturalis*, encompassing sonic phantom phenomena generated and experienced in nature, has become a significant component of the entire *Sonic Phantoms* compositional project. It is represented here by a selection of sound recordings and compositions, existing under the collective title of *Natural Phantoms* and exploring the phantom-producing polyphony of different natural environments. These sound pieces have been created from original source materials recorded in several field recording and listening expeditions that I carried out, in partnership with Francisco López, in rainforests and other natural environments in Brazil, Borneo, Cambodia, Australia, Bolivia, Chile, and South Africa between 2011 and 2019.

The sonic richness and complexity of these natural environments—a direct consequence of the abundance and diversity of their wildlife—provide an exceptional opportunity to experience and record the emergence of naturally occurring sonic phantoms. They have also been inspirational for the creation of transformative pieces from these recordings. The *Natural Phantoms* series of compositions is the result of a personal fascination with emergent natural sonic patterns, such as the multi-layered natural 'polyphony' of many wilderness environments, which I refer to as natural interlocking 'sonic tapestries.' There is also the captivating phenomenon of collective displays of 'acoustic synchrony'[5] (which we describe in more detail below) amongst frogs or insects; a phenomenon that relies upon the precise timing of central nervous system oscillators within these animals. Though much has been written about these kinds of fascinating tendencies towards synchronization distributed throughout the whole biosphere, the moments in which members of a natural 'chorus' go out of phase with one another are no less fascinating. The insistence of certain individual male frogs upon temporally separating their calls from others in a chorus, for example, has been

---

[4] Uexküll, *A Foray into the Worlds of Animals and Humans*.
[5] Walker, "Acoustic Synchrony."

Figure 6.2 Fieldwork location for environmental recordings and on-site listening in the rainforest of Borneo, Malaysia. Photograph by Francisco López, 2012.

taken as a proof[6] of their using calls as a means of fulfilling distinct group interaction functions: in this case, making themselves sonically stick out and thus increasing their chances of being noted and selected by females of the same species.

Before embarking on this journey in earnest, though, let us step back for a moment and survey the larger state of humans' creative sonic engagement with nature at present. It is unlikely that, in the near future, there will be any cessation of the discussion around

---

[6] Awbrey, "Social Interaction among Chorusing Pacific Tree Frogs, Hyla Regilla."

hybridization of so-called natural or organic environments and technical or synthetic ones, and the role this process plays in art and aesthetics (already within the humanities or 'post-humanities' there has been plenty of animated discussion around theoretical concepts such as Donna Haraway's 'nature cultures').[7] Certain technologically shaped urban environments (from machines to buildings to cultural communities), which represent relics or techno-archeological remnants of different stages in the industrial revolution era, are becoming scarce or distinctive enough to warrant documentation and preservation by audio recording, and have become as much a subject of attention for creative field-recording practices as their more untrammeled counterparts, as well as an annex to an expanding definition of 'ecology.'[8]

Another development, somehow paralleling that of approaching urban areas from the preservational perspective of a naturalist, is that of creating 'audio images' of the natural world that rely more upon high-tech or unusual equipment notable for its ability to dramatically alter our perception of these environments. This particular mediation of current sound-recording technology provides some very unusual and captivating listening perspectives that offer enormous potential for working from a musical or sound art perspective. The technical tools available today (multi-channel arrays, close-up microphones, hydrophones, contact piezo-disks, etc.) allow for a multitude of sonic perspectives that might often be surprisingly different from our normal sonic impression and memory of those environments. In doing so, these tools amply surpass their assumed role as representational devices, to become—especially when we have a musical or artistic vision—means of sonic exploration. This very topic was the subject of a roundtable discussion on technology held within the pages of *Scape*—the official journal of the acoustic ecology movement—that remains currently relevant in spite of its 2002 publication date. Within this larger conversation, Christine McCombe claims that "sound shapes experience, and technology allows me to shape sound in ways that I cannot begin to imagine or define,"[9] an endorsement of using technology within 'natural' environments for purposes beyond documentarian ones. Elsewhere, experimental musician John Hudak, in response to questions about the relation of technology to his work with acoustic ecology, claims that it not only speeds up his creative process but also "[provides] a world-wide presence"[10] that would otherwise be unavailable. Also within this realm of creative work with natural sound recordings, composer Barry Truax proposes that computer-controlled sound diffusion techniques can enhance spatiality.[11] This is a strategy that typically enhances the more

---

[7] 'Nature cultures' is a neologism coined by feminist technology theorist Donna Haraway, who uses it to describe the inseparability of the two supposedly opposing concepts embedded in this new term. In Haraway's reckoning, culture cannot be excised from nature since humans are a biological species, while nature is in itself a cultural construct. See Haraway, *The Companion Species Manifesto*.
[8] For examples, see Various, "Psychogeographical Dip CD | Discogs"; Various, "Régénération - Dégénérescence"; Various, "Siebzehn Bis ∞ (2015, File) | Discogs"; La Casa and Peyronnet, "Zones Portuaires."
[9] Quoted in Labelle and Brietsameter, "Questionnaire."
[10] Ibid.
[11] Truax, "Composition and Diffusion."

mundane simulative aspect of immersive sonic technologies that are commonplace today—and which have more commercial and considerably less interesting goals—such as IMAX theaters and VR headsets. Finally, perhaps the most radical departure and critical stance on recording as representation is probably that of Francisco López, who, from his double background as both composer and ecosystems biologist, offers a completely different understanding of the role of our recording machines, as well as our relationship to them and to 'reality' *through* them:

> When we hear what machines have heard and memorized, we might experience a revelation: the unfolding of the non-representational layers of sonic reality. Even more, the questioning of sonic 'reality' itself. In my view, this is the true, natural, and fruitful cooperation with machines of perception, particularly in their current state. Not the constant scorn of their 'limitations' to replicate that reality we seem to know so well, but rather our deep appreciation of what they have truly become as non-cognitive collaborators in our constant—perceptive, rational, aesthetic, spiritual...—quest in our interaction with reality; whether direct, referred, recorded or broadcasted.[12]

With my own personal perspective on such interactions between the experience of the 'natural' and the creative sonic work, field trips to diverse wilderness environments have become an integral part of my practice as an artist. Like many other artists working today with 'field recordings' in an iconoclastic, non-canonical way (e.g., Francisco López, Joe Colley, Lee Patterson, Israel Martínez, Eric LaCasa, Artificial Memory Trace), I am not interested in simply documenting an environment; unlike most standard strains of so-called 'acoustic ecology,' the primary emphasis for me is certainly not upon the diagnostic or indexical characteristics of sounds. Instead, I typically embark on a very personal, emotional, and artistic exploration that combines extended direct listening immersion with mediated exploration of the phantasmatic intricacies of these sonic environments. This approach reveals levels of sonic multiplicity and complexity that greatly surpass those of a standard sound impression or memory of a place.

When they are fundamentally informed by this on-site intensive listening engagement, recordings provide texturally rich and semantically open sonic materials on their own, which are often inherently ambiguous artifacts and which can be used as sources of 'sonic blotscapes.' Recordings that can particularly function as such are those naturally multi-layered with dense and complex sonic structures, characteristic of lush wilderness environments (even if this is obviously a very relative concept). They offer a rich potential to illuminate, transform, and accentuate any apparent or subsumed sonic phantoms in a variety of ways. Indeed, their further 'life in the studio'—particularly when this is not a simple representational one—is a renewed territory for

---

[12] López, "Music Dematerialized?"; Lopez et al., "Modes of Listening."

a differentiated, intriguing, and heightened listening experience; one that instills and inspires a compositional process born out of the sonic materials themselves.

For any of these levels of listening immediacy, sounds naturally emerging from wilderness environments can very often appear to be quite illusory, generating very direct and immediate sonic phantoms in our perception. Particularly when these environments have a spectrally broadband quality (i.e., with many simultaneous frequencies)—something not rare in nature—we often have the propensity to detect all kinds of aural 'meaningful' phantasmatic patterns embedded in them (perhaps the most common being the 'voices' we hear in the wind, in flowing water, in dense animal choruses). In consonance with Guthrie's view that animism itself is a by-product of perception from the workings of the brain itself ("voices murmur or whisper in wind and waves"),[13] I envision such naturally induced sonic phantoms as a kind of *sonic animism*.

6.2 Natural Polyphony

Beyond the usual perception of isolated animal calls, and also beyond the prototypical birdsong 'dawn choruses,' many natural sound environments pulsate with an astonishing variety of combinations of simple component structures that give rise to emergent sonic textures and larger-scale forms. Besides commonly appreciated qualities of 'beauty' or complexity of individual animal calls (which I recognize and appreciate), wilderness environments offer—perhaps less obviously—formidable collections of complex, self-organized structural patterns, which are often interestingly based on very uncomplicated individual elements.

Like hocketing instruments or voices, relatively simple patterns of alternation between sounding and silence of individual frogs or insects, for example, can typically give rise to the most densely interwoven textured polyphony. Between randomness and regularity, between species-specific features and the pressures of competition among calling neighbors, a multiplicity of parts—sometimes in astounding numbers—contribute to the whole, to the overall Gestalt we can identify as potential generator of sonic phantoms. This polyphony, therefore, does not refer to the simple presence of different non-human voices or animal calls in an environment (and this seems to me more of a metaphorical use of the term 'voice'), but rather attends specifically to the emergent character of rhythmic patterns, harmonies, sonic textures, and timbres. When we listen to them with such a musical ear, natural sound environments are composed of a multitude of simultaneous looping cycles, persistent repetitive structures, oscillations between sounding and silence, and continual shifts in their figure-to-ground motion, with sounds rising fleetingly to the foreground of our awareness and then receding back into the background flux.

---

[13] Animism, according to Guthrie, is the attribution of life to something where there is none. Guthrie, *Faces in the Clouds*.

6.2.1 Layering

The composer Henry Gwiazda relates a humorous anecdote about his attempt to create dense, spatially complex musical works using only a stereo speaker or headphone setup and software for virtual spatialization. While auditioning the software in question (Focal Point, designed for MacOS), he experienced the following:

> I found that I could make the sounds appear to move toward and away from me. The sounds were so vivid; I could almost see a bird flying over my head. The sound became visual. I removed everything in front of me—books, pictures—to better "see" the sound. I could not, however, place a bird sound at a low spatial location. The engineers I talked to explained that *we humans expect high sounds to be just that* [italics ours].[14]

This story betrays an interesting fact about human perception, namely our inclination to erroneously associate spatial characteristics with the pitch of sounds, even knowing what we know about creatures that can voice 'high' or 'low' tones irrespective of whether they spend their lives high in the trees or close to the ground. Our compositional habits continue to evince this, as countless musicians and sound artists still regularly use bands of high frequencies to indicate either a terrestrial or 'spiritual' flight, and low tones to suggest a corresponding identification with more intensely physical, 'ground-level' activities and modes of thinking. The natural world, to be sure, has its own ideas about tonal organization, which are not nearly as exclusively bound up in the kind of habitat where these creatures reside.

As a fundamental feature of the aforementioned polyphony, the multi-layered structure of natural environments has a clear natural expression in the partitioning of the frequency spectrum between different species or groups of species. Similar to the partitioning of the radio frequency spectrum among different stations, animal calls appear typically organized in different bands of their own frequency spectrum (including those frequencies that are beyond the upper limit of the human audible range: many creatures, such as bats, dolphins, and insects, communicate in an ultrasonic range that is well beyond what we can naturally perceive). With partitioning characteristics being first noticed among the more vocal species of frogs and birds, research on bioacoustic niches or acoustic partitioning continues to be submitted, and even the acoustic character of the underwater realm, with its distinct types of vocalization such as freshwater dolphins' echolocation, is beginning to be mapped to a greater degree than before (with thanks due in part to newer passive acoustic recording techniques, which "can be used at depths not accessible to humans, independent of weather conditions, and for a long term period").[15]

Bernie Krause, prominent acoustic ecologist and founder of the soundscape archive Wild Sanctuary, has done fieldwork that has been a significant driver of this

---

[14] Gwiazda, "Sound is Not Enough."
[15] Ruppé et al., "Environmental Constraints Drive the Partitioning of the Soundscape in Fishes."

understanding of bioacoustics. In recent years his theory of 'acoustic niches'[16] or partitioning has received some empirical validation with the help of Almo Farina and Rachele Malavasi (who determined that birds returning from annual migrations would also go through an annual process of re-learning their 'niches' after an initial period of inter-species cacophony).[17] In acoustically dense natural environments, where there exist any number of distinct voiceprints or (to borrow Krause's own term) 'soundmarks' identifying the creatures within, there will be strong pressure on the sound-making species to maximize the effectiveness of their calls by partitioning them within the acoustic landscape. Purposive communication on both intra-species and inter-species levels requires animals to have their own acoustic niche and thus minimize the degree of interference that would be brought about by masking. In order to be heard, within specific ranges of possibilities for biological mechanisms to produce (and hear) sound, each species has evolved and adapted its own acoustic stream or niche within the spectral structure of any natural environment.

As is in fact prototypical in many rainforest environments, I encountered an exceptionally prime example of this sonic partitioning in one of the night recordings I did in the Borneo rainforest. The recordings in question are representative of a typical late night in that forest, insofar as there is a clear division between the two main groups of animals that fill up the sonic environment, in terms of the organization of sound frequency ranges: several species of frogs in the lower part, between approximately 700 Hz and 4 kHz, and different species of insects (crickets, cicadas, grasshoppers, katydids), from 4 kHz all the way up to the upper limit of the human audible range. In this case, said range was also coincidental with the limit of the recording range of the microphone equipment used. Within these two main sectors, individual species or sub-species are in turn further differentiated with their specific frequency bands, sometimes very narrowly, offering a spectral appearance of sharply defined, thin horizontal strips.

6.2.2 Interlocking

In addition to the spectral organization, natural polyphony is also built from an acoustic temporal partitioning. As can be clearly seen in figures 6.3 and 6.4 below, some species produce a perceptibly continuous stream of sound (mostly insects), while others emit calls that are clearly fragmented or interjected with silences. In many natural environments, from the perennially humid, like rainforests, to the seasonally and dramatically contrasting dry–wet, like arid ecosystems, the most common and predominant of such calls are the sounds of frogs and toads: they whistle, toot, and hoot, making interlocking calls that are short in duration and voiced separately. Given the considerable overlap in their frequency ranges, their naturally selected vocalizing strategy, as mentioned before, is to alternate their sounds with those of potential rivals, and this is what gives rise to their intricate interlocked polyrhythmic textures. The

---

[16] Krause, "The Niche Hypothesis."
[17] Malavasi and Farina, "Neighbours' Talk"; Farina et al., "Avian Soundscapes and Cognitive Landscapes."

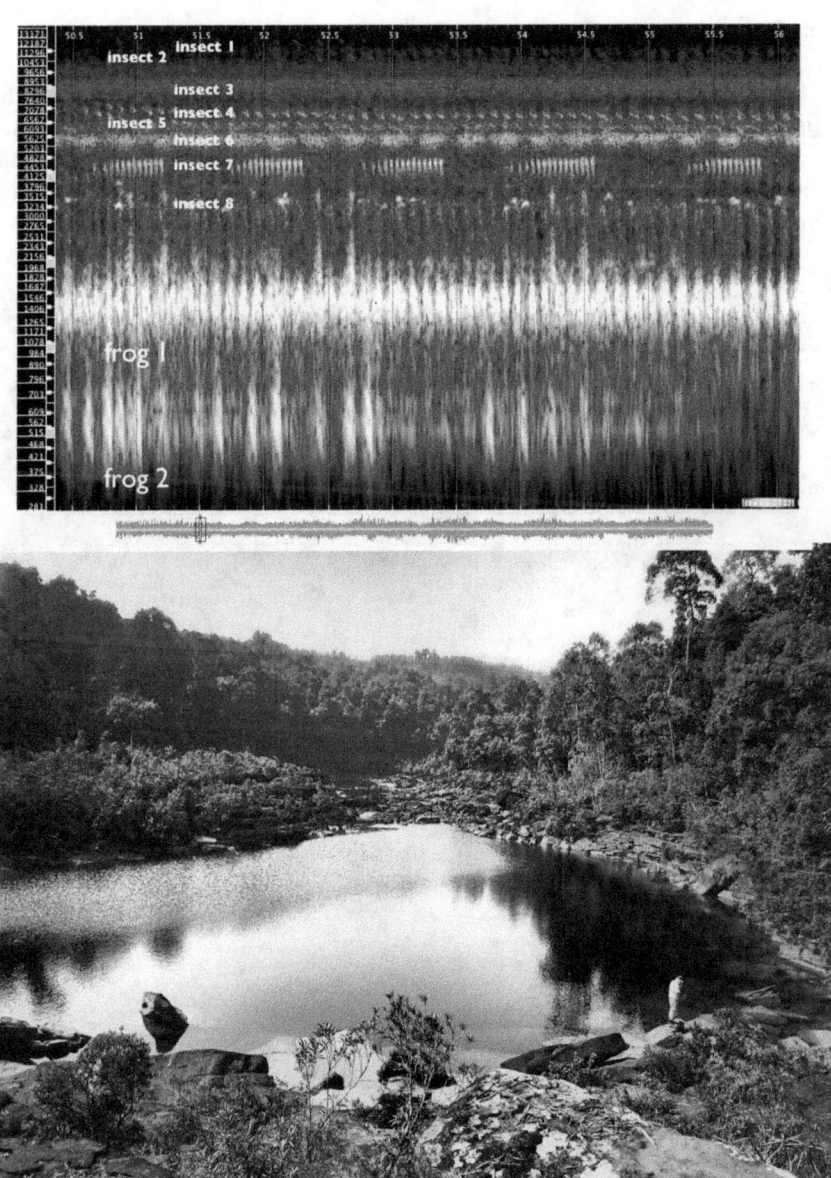

Figure 6.3 Interlocking patterns of frog and insect sounds as shown in a spectrogram (top image) from a prototypical example of the nocturnal sonic environment of the Cardamom Mountains rainforest in Cambodia (bottom image). The detailed inspection of the spectrogram, along with the audio recording, permits the recognition of at least eight different species of unidentified insects and two species of unidentified frogs. X-axis: time, with 500 ms vertical divisions; y-axis: frequency, logarithmic scale, 280 Hz to 13k Hz. Original recordings and images by Barbara Ellison, 2013–2014.

complex overlapping and interlocking rhythmic patterns allow both each individual frog and the entire emergent chorus to be heard simultaneously.

In the constantly changing natural sonic environments, sound events have their structured timeline, but rarely synchronize or 'keep together in time.'[18] The phenomenon of perceiving and synchronizing to a guiding pulse or beat, and thus coordinating actions, is rare enough among animals that it is generally considered to be unique to Homo sapiens.[19] Albeit rare, some animals, such as frogs and insects, do demonstrate

Figure 6.4 Fieldwork locations for environmental recordings and on-site listening in the savanna environment of Mmabolela Reserve, northeastern South Africa (top image). Spectrogram from a prototypical nocturnal recording. X-axis: time, with half-second vertical divisions; y-axis: frequency; linear scale, range 45 Hz to 23.5 kHz, slightly above human audible range (bottom image). This spectrogram shows the conspicuously layered structure with most of the frequency bands occupied by different kinds of insects (brighter horizontal lines; in increasing frequency: crickets, grasshoppers, and katydids). Photographs, recording, and spectrogram by Barbara Ellison, 2015.

[18] McNeill, *Keeping Together in Time*.
[19] Wallin et al., *The Origins of Music*; Greenfield and Schul, "Mechanisms and Evolution of Synchronous Chorusing."

an intricate combination of acoustic alternation and synchronous behavior, which biologists have termed 'acoustic synchrony' or 'synchronous chorusing.'[20] This term refers not just to signals in unison, but rather to a complex mixture of synchronous and precisely asynchronous signals. Collective patterns generated by large groups of frogs and insects show a sophisticated resulting structure with correlation in both speed and phase of the rhythms of neighboring signalers.[21] This is considered partly an epiphenomenon and partly a consequence of some individual reactions, with multiple simultaneous combinations of both synchronized and alternating calls, from neighboring competing individuals.

From an artistic or musical perspective, these collective interlocking patterns, with their polyrhythmic structures, appear as highly structured and densely interwoven sound environments; striking examples of what many would consider—more often than not in a blatantly anthropomorphic fashion—as 'natural compositions.' Given the increasing prevalence of 'graphic scores' mentioned earlier, as well as the recent history of dedicated software applications that translate color gradients into varying qualities of pitch, loudness, and timbre, many musicians and composers appear to find it difficult to resist the dangerous temptation of seeing spectrograms with interlocking patterns, like the ones in the figures above, as directly equivalent to 'scores.' Indeed, Krause—who is no stranger to the work of R. Murray Schafer—draws parallels between the visual appearance of the spectrographic readouts he had first seen in the 1980s, and Schafer's own ventures into graphic notation.

Without going into the age-long debate of whether or not these, or any other natural, non-anthropogenic sound constructions are music, or in what sense or under which conditions they could be music,[22] my perception and my approach in working with them is certainly musical in a sense that most would recognize as such. These natural interlocking patterns have become the direct source or the structural and harmonic inspiration of my *Natural Phantoms* compositions (as well as my compositional attitude in general), in some cases with very little or no transformation at all of the original patterns. In fact, in a very significant way, my interest and appreciation for natural sonic phantoms is somehow the antithesis of the analytical perspective represented by the 'niche' division described above: it is precisely the blurring of discrete separation and recognition, the lack of clear boundaries, the confusion between signals, the absence of individuated indexical sonic elements, that is at the core of the generation and existence of these evasive and ephemeral phantasmatic manifestations. And who is to say that that is *not* the way an ecosystem, and all its integrating sonic creatures, are 'hearing themselves'?[23]

There is also an additional inherent component of the natural interlocking patterns that can only be fully appreciated for its sonic or musical characteristics upon deeper scrutiny: the micro-timbral structure of the apparently simple individual calls.

---

[20] Walker, "Acoustic Synchrony."
[21] Greenfield and Schul, "Mechanisms and Evolution of Synchronous Chorusing."
[22] See López, "Environmental Sound Matter"; Rothenberg, *Why Birds Sing*; Weiss, *Varieties of Audio Mimesis*.
[23] López, "Sonic Creatures."

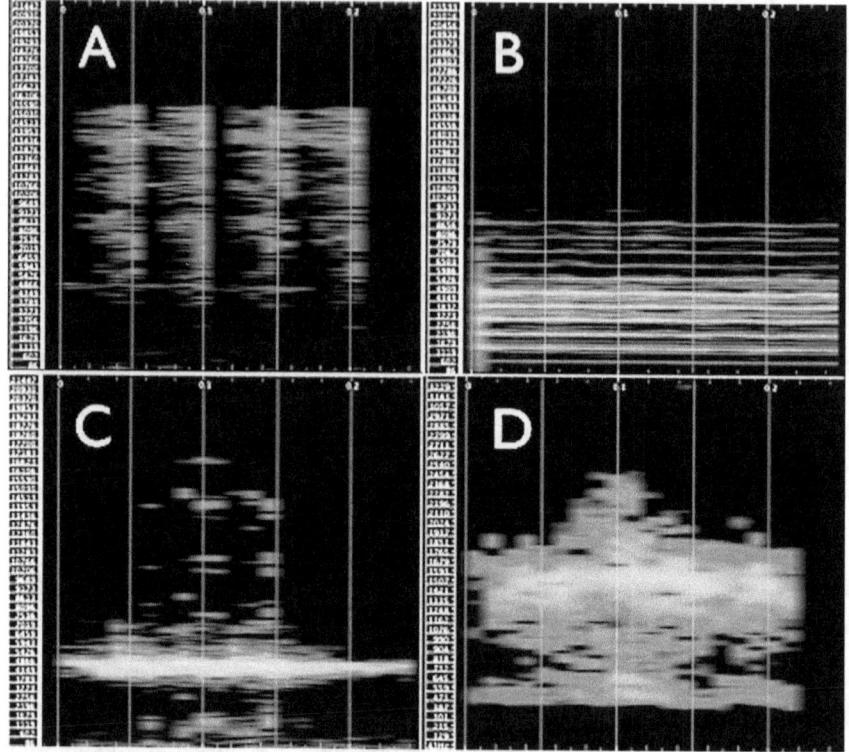

Figure 6.5 Spectrograms with prototypical examples of micro-timbral structure in several isolated individual animal calls (from my own field recordings in Borneo, Malaysia; Mamori Lake, Amazon, Brazil; and Cardamom rainforest, Cambodia). A: grasshopper; B: cicada; C: cricket; D: frog. X-axis: time, with 50 ms vertical divisions; y-axis: frequency, linear scale; all insects shown within the human audible range, 20 Hz to 20 kHz; frog shown within the range 20 Hz to 3 kHz.

As would happen with the short attacks characteristic of percussive instruments, a detailed spectral observation of isolated calls reveals the micro-structural complexity that produces their timbre or tone.

A closer inspection at this range of temporal scales shows how micro-temporal sound-producing mechanisms such as stridulation (rubbing together body parts; in crickets and grasshoppers), tymbalization[24] (resonance of body cavities with ultra-fast muscle contractions; in cicadas), and periodic amplitude modulation (vibration of the glottis membrane; in frogs and toads), result in the astonishing variety of acoustic signals produced by insects and frogs. To wit, whistles, clicks, croaks, chirps, trills, emissions that seem to come from 'electronic' tone generators or noise generators,

---

[24] Sound-producing system in cicadas consists of two coupled resonators: firstly, the sound-producing tymbal, and secondly, the Helmholtz resonator that consists of the abdominal air sac and tympana. The tymbals produce the sounds, the resonator radiates and amplifies the sounds produced. See Bennet-Clark and Young, "A Model of the Mechanism of Sound Production in Cicadas."

and more besides. This micro-timbral component becomes audible and more relevant in the case of integrating natural recordings into larger compositions, if any form of sound processing is used—as I did in some of my *Natural Phantoms* compositions—that will pitch down (frequency transpose) or time-stretch small sections of recordings with interlocking patterns.

6.2.3 Transitions

While the detailed timbral structure of animal calls can be seen as a micro-temporal level of natural polyphony (on the scale of fractions of a second), and whilst both layering and interlocking patterns would constitute its meso-temporal structural features (best appreciated at ranges between seconds and a few minutes), there is yet another scale that would represent a relative macro-temporal level of natural polyphony: the slow transitions and crossfades among different species caused by their periods of activity, which take place over periods of many minutes to hours.

All natural environments have these activity transitions—daily sequences of temporal niches for different species—with their corresponding sonic manifestations. They are predominantly slow-paced, with intricate, progressive, morphing transformations, and can only be appreciated at temporal scales that largely exceed any standard experience of attentive listening or recording. Using the analogy of a popular audio-recording concept, these transitions are ever-changing 'mega-mixes' of shifting, layering, and interlocking patterns, which slowly transform during and in between the different phases of the daily rhythms. From dawn to dusk and overnight until the cycle repeats again, this dynamic circadian polyphony will gradually change between the seasons in each given environment.

From the different natural environments that I have had the opportunity to experience and record, one of the richest, most complex and dramatic examples of these macro-temporal patterns is the dusk transition of the Borneo rainforests. It kicks off every evening with incredible precision at six o'clock sharp, with the spectacularly loud, repetitive calls of the empress 'trumpet' cicada (*Megapomponia imperatoria*), which is also known locally, for obvious reasons, as the 'six o'clock cicada.' The first thirty to forty-five minutes or so are dominated by the sounds of different cicadas.[25]

Around seven o'clock in the evening, the activity of the cicadas cross-fades into an intense, lengthy crescendo generated by different layers as represented by different insect species (crickets, grasshoppers, katydids) all across the higher part of the frequency spectrum, later given additional modifications by the incorporation of frogs. It constitutes a continual *fortissimo* broadband sonic broadcast. Textures and sound events, which literally strike the listener as being like multiple-layered tracks in a large multi-track sound composition, fade in and out and are switched on and off, according to a precise time-line for the dusk transition. Meanwhile, sounds are moving all around in audio space (many insects, including the 'trumpet' cicadas, fly

---

[25] For a detailed description of the circadian rhythms of cicadas in this ecosystem, see Gogala and Riede, "Time Sharing of Song Activity by Cicadas."

from tree to tree) and there is vertical stratification, as most cicadas and katydids sing high up in the canopy whereas frogs and crickets are singing mostly at ground level. Sonically, it is an overwhelmingly powerful experience and one that induces sonic phantoms of the most extraordinary nature. It is a feat that owes itself to the astonishing diversity of the sounds on tap: the pulsating, beating, hocketing, droning, buzzing, and tooting of the dusk 'performers.'

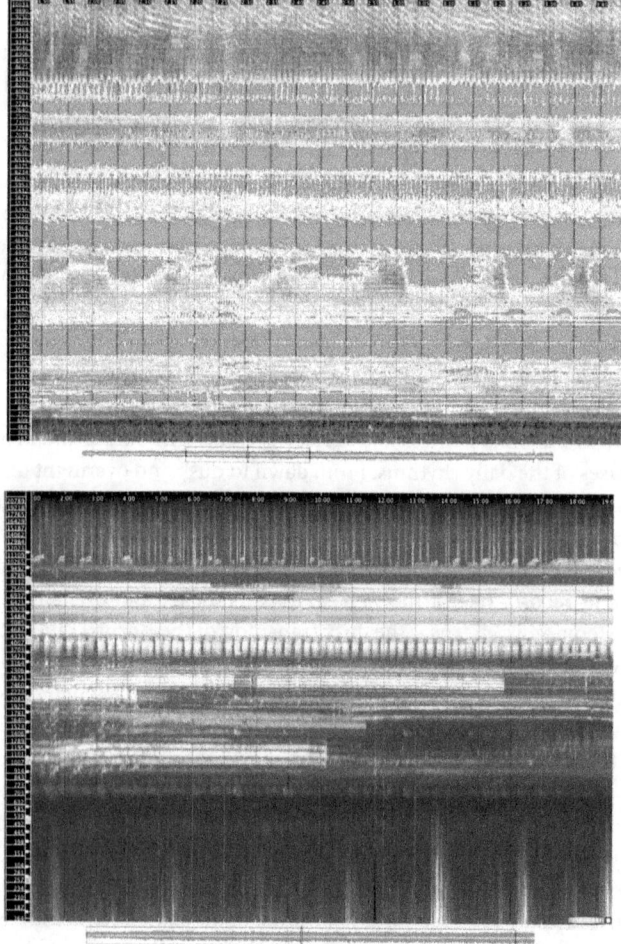

Figure 6.6 Two spectrograms showing a prototypical example of the multilayered rainforest dusk transition (from Borneo, Malaysia) at different scales. Virtually all sounds are produced by loud and persistent insect calls (crickets, cicadas, and katydids; brighter horizontal bands in the image). Top image: x-axis: time, with 5-second vertical divisions; y-axis: frequency, linear scale, 200 Hz to 10 kHz. Bottom image: x-axis: time, with 1-minute vertical divisions; y-axis: frequency, linear scale, 200 Hz to 21 kHz. Original recordings and images by Barbara Ellison, 2012–2013.

Everything in the observable environment becomes mixed into this very densely layered performance. Layers are segregated and juxtaposed in time, sometimes by surprisingly violent contrast, sometimes with more subtly overlapping effects: fading in and out, fusing together and then splitting apart to stand out momentarily in the foreground, then just as quickly receding to a background plane, overlapped or masked by a another sound; undergoing a slow metamorphosis from one state to another.

6.3 *Natural phantoms*

The compositions of the *Natural Phantoms* series have been directly inspired by the listening and recording experience in the wilderness environments described above. Moreover, in terms of source materials, all of them were created using the field recordings made on those trips as the sole sonic elements for the pieces. In spite of considerable amounts of subsequent studio work, all of these compositions are pervaded by a sense of organicism that results from an ambitious attempt at clearly reflecting the 'feel' of that natural polyphony by keeping a significant proportion of its natural structures and timbral features intact.

In the process of compiling the *Natural Phantoms* series, I have worked in a very intuitive way, letting the materials—and my listening memory of the environments from which they originated—guide a process of semi-automatic composition in terms of editing, processing, enhancing, accentuating, combining, mixing, and so on. I perceive this as a subtle intervention—at times even rising to the level of an 'interaction'—that has always tried to be sensitive and responsive to the sonic nature of the materials and their structure. All the fundamental features of the natural polyphony—micro-timbre, layering, interlocking, transitions—have a natural 'sonic phantasmatic-inducing' power, exerted through their characteristics of repetition, persistence, noise, and the myriad sonic blends of these different elements. My search for and induction of sonic phantoms in this case has been therefore particularly dependent upon the actual natural materials and their organization; the very compositional framework and the development of the *Natural Phantoms* pieces derive from those natural features.

In this series of compositions, I have considered and explored all the temporal levels mentioned above: micro-, meso-, and macro-temporal. They also encompass a wide spectrum of studio electroacoustic intervention on the actual materials, from 'straight' field recordings unmodified by additional editing, to heavily processed or re-organized recordings. The *Natural Phantoms* pieces, made of recordings from Borneo, Cambodia, and Australia, represent examples of the first category, with relatively subtle accentuation work carried out to enhance and heighten the inherent phantasmatic nature of the sounds: the techniques utilized to achieve this end included an elaborately detailed equalization and occasional, selective gain enveloping. At the opposing end of the spectrum would be my first composition of the series, *Natural Phantoms #1*, with a significant use of transpositions that bring out micro-timbral elements, turning them into structural features.

Combining several simultaneous strategies, another composition of the series entitled *Natural Phantoms #10*, on its part, uses extreme accentuation and transposition of individual layers that are subsequently combined with their non-transposed counterparts in variable combinations, as well as mixed and organized in time following an inspiration from natural transitions between layers.

Finally, *Natural Phantoms #9*, created from underwater recordings I carried out in the Limpopo river at Mmabolela Reserve in South Africa, was composed and constructed with extensive editing, giving rise to a multitude of fragments that were re-combined and mixed in a 38-channel structure. Many of the original sounds—all of them from fish and underwater insects—were kept intact in this structure, along with their natural interlocking organization. Others were modified by different transpositions, and new interlocking and transitional formations were engineered from the intricate combination of all the layers. A further element for the induction of sonic phantoms in this piece was the deliberate raising of the noise level in some of the recordings.

6.4 Beyond Composition

Besides and beyond that which is only apparently 'compositional,' the profoundly engaging experience of listening and interacting creatively with these examples of *Phantasma Naturalis* might also implicitly suggest much more about the significance of naturally derived sonic phantoms, the sound of the biosphere as a whole, and about our own situation within that complex and fascinating 'natural' sonic-aural microcosm. Among the many authors and thinkers to make the biosphere a focus of the work is Marc Changizi, whose book *Harnessed* provides a kind of twenty-first century update to the previously mentioned musical philosophy of the Abbé du Bos: the book lays out an intriguing argument for the essentiality of the natural world's influence upon more distinctly human sound environments, at both micro and macro levels. For Changizi, the two auditory capabilities that we most pride ourselves in (i.e., speech and music) are capabilities that "evolved over time to be simulacra of nature."[26] While his thesis makes clear that nature mimicry was never the *sole* purpose of communicative tools that were, after all, meant for "transmitting thoughts to others" and "evoking feelings in others,"[27] respectively, his work follows that of aforementioned researchers like Rothenberg and Krause in its skepticism towards the supposed human exclusivity of musical composition.

More important than this, though, is how the attempt to create 'simulacra of nature' relates to the relentless drive towards ascribing meaning that was discussed early on in this story. In the same way that the study of the sonic phantoms associated with EVP

---

[26] Changizi, *Harnessed*.
[27] Ibid.

and OVP can provide valuable insights into certain types of human universality (i.e., our tendency towards personalized, emotionally charged interpretations of extreme perceptive ambiguity), so too can the creative practice with *Phantasma Naturalis* deliver us a profound realization of our universal inextricability from nature. The musical and compositional techniques we have adapted from nature may no longer be bound up with ensuring survival (our use of the hocket, for example, seemingly has much more to do with purposeless play than it does with the utilitarian maintenance of acoustic niches). However, those techniques are never completely separate from the organizational logic of the natural environment, and even the 'purposeless play' mentioned here might as well pre-date human culture.[28] When we listen for sonic phantoms in natural sound environments, and apply to them the same process of semantization common to all other types of sonic phantoms, we are reacquainting ourselves with one of the fundamental processes that—for better or worse—gave us our distinct role vis-à-vis the rest of the biosphere. We could argue that the need to ascribe meaning to nature, and thus view nature as being in a fairly constant and *direct* communication with us, has been a prerequisite for our eventual growth as cultural beings. The hunt for 'natural' sonic phantoms thus gives us an unequivocal reminder of this fact, while allowing this archaic process to continue anew.

---

[28] "Play is older than culture, for culture, however inadequately defined, always presupposes human society, and animals have not waited for man to teach them their playing." Huizinga, *Homo Ludens*.

Figure 7.1 *The Phantom Within.*

7

Coda

Otoacoustic Emissions: The Phantom Within

Having thoroughly examined how sonic phantoms might arise from each of the primordial Boethian realms, it is perhaps worth asking at this concluding stage: is there any other phantasmatic territory too ambiguous to be clearly ascribed to any of the above? That is to say, are there 'phantoms' that retain, for instance, some aspects of human vocalization and yet lack the human agency common to the other examples investigated? Or are there manifestations of sound that impress upon us as if being produced by an 'instrument,' yet also require the definition of instrumentation to stretch our existing musical taxonomies of such? In the spirit of the very ambiguity that brings sonic phantoms to life in the first place, we would like to finish this long-range exploration with a final phenomenological sketch that, if it achieves nothing else, will reinforce how perception of the sonic phantoms can truly border on the universal, and simply be closer to us than we ever expect.

The psychoacoustic composer Maryanne Amacher (1938–2009), though mentioned in passing earlier, merits an additional, more prominent appearance in this text before it comes to its close. She is a largely unique case study in phantasmatic sound; one of the most fascinating aspects of her creative research has, curiously, not been embraced by more than a handful of sonic explorers[1] in a world increasingly saturated with both music and raw experimentation with aurality (some notable exceptions will be discussed momentarily). This particular enduring piece of her legacy is her creative utilization of the phenomenon discovered by David Kemp in 1978 and scientifically known as 'otoacoustic emissions' (hereinafter OAEs). Known alternately as 'Kemp tones,' and personally described by Amacher as 'response tones,' these acted as a phantasmatic percept that provided her with her signature departure from conventional music composition. She nurtured a particular interest in what psychoacoustic researcher Juan Roederer referred to as 'first order superpositions,' or sonic overlays that would cause the listener to respond "not to the real world instruments, but to certain EXTREMELY RESONANT INSTRUMENTS

---

[1] Besides Amacher, to our knowledge, Phill Niblock, Zbigniew Karkowski, Jakob Kirkegaard, Thomas Ankersmit. See also Kirk, "Otoacoustic Emissions as a Compositional Tool" and Connolly, "The Inner Ear as a Musical Instrument."

[capitals in the original] within the anatomical structure of the ear."[2] As Amacher hints here, OAEs are a category of sound in which the cochlea produces acoustic energy, utilizing the outer hair cells to provide the mechanical energy that then acts upon the basilar membrane. Their discovery has had profound implications for the field of neurology, but with more obvious implications for the field of psychoacoustics: Rudolf Probst, for example, posits that "the discovery of OAEs permitted the proposition of testable hypotheses to account for several psychoacoustic phenomena, including the microstructures of behavioral sensitivity and loudness enhancement, which were incomprehensible in terms of conventional models of the auditory system."[3]

Already, astute readers may be drawing sonic comparisons to a well-known pathological condition that behaves in such a way, and so it is important to clearly distinguish OAEs from sounds produced by tinnitus (or similar irritants like the so-called 'Taos hum' or 'Kokomo hum' that affects roughly 2 percent of the global hearing population). Though tinnitus and OAEs bear some perceptual similarities (and 50 percent or so of the population capable of producing spontaneous OAEs *will* experience them as a kind of tinnitus),[4] they still differ greatly in many salient aspects. For one, only OAEs are actually physically detectable and (more importantly) recordable, provided one has a small enough microphone to fit directly into the ear canal. They can be monitored and recorded using condenser microphones attached with tubes to the ear canal, though with the caveat that they are best captured while the human subject is anesthetized (otherwise frictional noise of various kinds may be at too disruptive a level).

Amacher was careful to distinguish between these 'response tones' and a subtler, more distinctly phantasmatic type of 'second order' superpositions, in which a sort of uncanny 'listening to listening' was achieved. We could argue that this is another point that OAEs share with previously mentioned varieties of sonic phantoms: they provide an illusion of an external presence that, when we come to understand its true origins and mechanics, imparts a kind of semantic information about the true extent of our psycho-physical functioning. These, she stressed, were not "present in the cochlear fluid at all" but within the neural interactions of the brain, a fact which essentially meant that we "hear an *evolved sensitivity* [italics in the original], extracting information on details of the vibration pattern."[5] From here, Amacher's concept of a 'perceptual geography' becomes less notional and more of a concrete reality, as she was able to conceive of the ear and brain as "auditory dimensions in their own right, rather than *adjuncts* of an acoustic space."[6]

---

[2] Amacher, "Psychoacoustic Phenomena in Musical Composition."
[3] Probst, Lonsbury-Martin, and Martin, "A Review of Otoacoustic Emissions."
[4] Bauman, *Phantom Voices, Ethereal Music & Other Spooky Sounds.*
[5] Amacher, "Psychoacoustic Phenomena in Musical Composition."
[6] Ibid. See also Amacher, *Sound Characters Making Sonic Spaces.*

In order to then bring OAEs into more of a compositional framework—particularly for the pieces exhibited on her 1999 *Sound Characters* CD compilation on the record label Tzadik—Amacher cleverly utilized other techniques from the aforementioned repertoire to allow listeners to better realize a personal 'otoacoustic performance.' The opening piece on this album, *Head Rhythm 1/Plaything 2*, begins with a sort of bracing, high-frequency hocket of pure electronic tones that, when heard on stereo headphones, is remarkably effective in immediately setting a stage wherein listeners must question whether they are the producers or receivers of these tones. This is paired with an overlapping series of heavily effected, oceanic drones that bring to mind previous observations made in this text about the phantasmatic abilities of persistence (the latter effect is more thoroughly explored in the 22-minute track *Synaptic Island (Vms 3)*).

The Danish-born Jacob Kirkegaard is another artist and composer whose repertoire, like Amacher's, often focuses on revealing previously unheard acoustic phenomena—or, more accurately, that which was previously assumed to be inaudible. Kirkegaard then uses this newfound material as a springboard to a greater breadth of sensory and mental experience overall. His work deserves some special mention within this narrative, for his 2007 *Labyrinthitis* installation and recording has taken the concept of OAEs a step further than what has just been discussed here. Kirkegaard's ambition has been to place a magnifying glass onto this most private form of human 'singing,' as Amacher has done, but to do so in a way that triggers a sort of chain reaction: in other words, *Labyrinthitis* is novel for its attempt to use OAEs as actual recorded material meant to generate OAE responses in an *audience* rather than solely in the headspace of the 'emitter.' As with Amacher's work, *Labyrinthitis* relies upon other 'phantom-generating' techniques such as persistence and layering: the overall effect of hearing this on a home listening format is not unlike the effect achieved from, say, one of the drone-scapes of Eliane Radigue or Charlemagne Palestine, with the spectral character of the sounds initially bringing to mind tinnitus but then flourishing into timbral qualities that are uncannily like those of an organ.

To this end, and with the help of Professor Torsten Dau from the Center for Applied Hearing Research in Denmark, Kirkegaard was able to make a distinction between OAEs depending upon the way in which they are elicited: there are those that arise spontaneously (SOAEs) and the type Amacher has put to use in her compositions—that is, those that arise as a result of direct sonic stimulus. The latter are more properly referred to as stimulus-frequency otoacoustic emissions (SFOAEs), or alternately 'synchronously evoked otoacoustic emissions,' which can be generated by, per Probst, "a pure tone that is, in most instances, swept relatively slowly over a specific frequency range."[7] Kirkegaard's experiment with this material was temporarily stymied when he realized, upon examination, that he did not in fact emit SOAEs himself (notably, younger women are the demographic most likely to experience such). However, the disappointment was moderated by the fact that distortion-product otoacoustic emissions (a.k.a. DPOAEs, which are regularly used to determine the health of infants'

---

[7] Probst, Lonsbury-Martin, and Martin, "A Review of Otoacoustic Emissions."

hearing) could still be coaxed from him if his hair cells were "stimulated in certain ways that make them emit tones in response,"[8] a process that eventually revealed a tonal ratio of 1:1.2 as being the most effective for coaxing out these tones. Kirkegaard describes his 'eureka' moment as follows:

> As the speaker played the two pure tones of the above mentioned ratio in my ears...I could hear three tones: those two tones produced by the tiny speaker and my own evoked otoacoustic emission. I heard my own ear play. And since it played in response to the incoming sounds—which I could also hear—I could hear myself hearing.[9]

This sort of 'meta-hearing' process described here is maybe the most clearly OAEs come to the realm of the phantasmatic: unlike the other types of OAE mentioned here, DPOAEs can elicit tones that do not merely mimic the eliciting stimuli, having completely different pitch/frequency characteristics. Forms of layering (or, more accurately, 'intermodulation' deriving from the use of a bitonal stimulus) as well as 'expanding through narrowing' all seem to be at work here. It is fascinating to consider, again, that while we are engaging in the sort of active listening that has been described earlier to apprehend all kinds of sonic phantoms, here we have a form of listening in which we externalize the results of our phantasmatic voyage.

All told, OAEs remain a captivating subject of study and experimentation for those interested in the phantasmatic, for all the reasons that artists like Amacher and Kirkegaard have chosen to utilize them in their work, and also for the fact that their story is very much an ongoing one.[10] Researcher Franz Frosch reminds the reader that "the mechanisms of their generation are still not well understood,"[11] and while multiple working hypotheses exist regarding preconditions of their *production* (e.g., "spontaneous mechanical oscillations within the cochlea, and perhaps...motile properties of the outer hair cells"),[12] there is less available research on the circumstances that may surround the wide variance in tonal and musical qualities of these emissions. As this research gradually accumulates, though, we can still share in the wonderment that comes from having our own ears become 'transmitters' of the phantasmatic.

It seems fitting that, having already traveled so far afield, we would end this odyssey by returning to the most intimate means of attaining phenomenological knowledge: dramatically re-examining the very apparatus we use for interpreting the audible world. Whether one finds the existence of something like OAEs to be comforting or distressing (or something else entirely), their presence provides one

---

[8] Probst, Lonsbury-Martin, and Martin, "A Review of Otoacoustic Emissions."
[9] Kirkegaard and Museet for Samtidskunst, *Earside Out*.
[10] Connolly, "The Inner Ear as a Musical Instrument"; Kirk, "Otoacoustic Emissions as a Compositional Tool."
[11] Frosch, "Hum and Otoacoustic Emissions May Arise Out of the Same Mechanisms."
[12] Ibid.

final affirmation of the sonic phantasmatic and the universal attainability thereof. We contend that the world is richer for this fact, and that a growing acknowledgment of our innate drive towards unique and personally resonant meanings—whether as 'primary' or 'secondary' composers—points towards whole new disciplines of creative expression and interaction, and further refutations of the cynic's shopworn refrain that 'everything has already been done.'

Bibliography

Alcock, James. "Electronic Voice Phenomena: Voices of the Dead? | Skeptical Inquirer." *Skeptical Inquirer*, 2004. https://skepticalinquirer.org/exclusive/electronic_voice_phenomena_voices_of_the_dead/. Accessed October 24, 2019.

Aldridge, David, and Jörg Fachner. *Music and Altered States Consciousness, Transcendence, Therapy, and Addictions*. London; Philadelphia: J. Kingsley, 2006.

Amacher, Maryanne. "Music of the Spheres: Reflections on Karlheinz Stockhausen." *Artforum International* 46, no. 7 (2008): 302.

Amacher, Maryanne. "Psychoacoustic Phenomena in Musical Composition." In *Arcana III : Musicians on Music*, edited by John Zorn, 9–24. New York: Hips Road, 2008.

Amacher, Maryanne. *Sound Characters Making Sonic Spaces*. New York: Tzadik, 2008.

Arnheim, Rudolf. *Art and Visual Perception: A Psychology of the Creative Eye*. Berkeley: University of California Press, 1974.

Arom, Simha, Martin Thom, Barbara Tuckett, Raymond Boyd, and Gyorgy Ligeti. *African Polyphony and Polyrhythm Musical Structure and Methodology*. Cambridge University Press, 1991.

Ashley, Robert. *Automatic Writing* [1996, CD]. Lovely Music, 1996.

Austin, L. "Sound Diffusion in Composition and Performance: An Interview with Denis Smalley." *Computer Music Journal* 24 (2000): 10–21.

Awbrey, Frank T. "Social Interaction among Chorusing Pacific Tree Frogs, Hyla Regilla." *Copeia* 1978, no. 2 (1978): 208–214.

Bailey, Thomas Bey William. *Micro Bionic Radical Electronic Music & Sound Art in the 21st Century*. Creation Books, 2009.

Bailey, Thomas Bey William. *Unofficial Release: Self-Released and Handmade Audio in Post-Industrial Society*. Belsona Books, 2012.

Bailey, Thomas Bey William. *To Hear the World with New Eyes: A Cultural History of the Synesthetic and Cross-Sensory Arts*. Edited by Francisco López. Murcia, Spain: SONM (Sound Archive of Experimental Music and Sound Art), Puertas de Castilla Center, 2016.

Baker, Phil. *The Book of Pleasure (Self-Love): The Psychology of Ecstasy*. London: The Author, 1913.

Baker, Robert A. *They Call It Hypnosis*. Buffalo, NY: Prometheus Books, 1990.

Ball, Hugo, and John Elderfield. *Flight Out of Time: A Dada Diary*. Berkeley: University of California Press, 1996.

Banks, Joe. "Rorschach Audio: Ghost Voices and Perceptual Creativity." *Leonardo Music Journal* 11, no. 1 (2001): 77–83.

Bateson, Gregory. *Steps to an Ecology of Mind*. Chicago: University of Chicago Press, 2000.

Bauman, Neil G. *Ototoxic Drugs Exposed: Prescription Drugs and Other Chemicals That Can (and Do) Damage Your Ears*. Stewartstown, PA: GuidePost Publications, 2002.

Bauman, Neil G. *Phantom Voices, Ethereal Music & Other Spooky Sounds: Musical Ear Syndrome: Unraveling the Mysteries of the Auditory Hallucinations Many

Hard of Hearing People Secretly Experience. Stewartstown, PA: Integrity First Publications, 2011.

Bauman, Neil G. *When Your Ears Ring!: Cope with Your Tinnitus—Here's How.* Stewartstown, PA: Integrity First Publications, 2013.

Beardslee, David C., Michael Wertheimer, and Edgar Rubin. *Readings in Perception.* Princeton, NJ: Van Nostrand, 1960.

Beaudry, Nicole. "Singing, Laughing, and Playing: Three Examples from the Inuit, Dene, and Yupik Traditions." *The Canadian Journal of Native Studies* 8, no. 2 (1988): 275–290.

Beaudry, Nicole. *Toward Transcription and Analysis of Inuit Throat-Games: Macro-Structure.* 1990.

Becker, Judith. "Music and Trance." *Leonardo Music Journal* 4 (1994): 41–51.

Beheydt, Ludo. "The Semantization of Vocabulary in Foreign Language Learning." *System* 15, no. 1 (1987): 55–67.

Bennet-Clark, H. C., and D. Young. "A Model of the Mechanism of Sound Production in Cicadas." *Journal of Experimental Biology* 173 (1992): 123.

Berendt, Joachim Ernst. *Nada Brahma: The World is Sound: Music and the Landscape of Consciousness.* London: East West, 1988.

Berliner, Paul F. *The Soul of Mbira: Music and Traditions of the Shona People of Zimbabwe: With an Appendix Building and Playing a Shona Karimba.* Chicago: University of Chicago Press, 2007.

Black, S. R. "Semantic Satiation and Lexical Ambiguity Resolution." *The American Journal of Psychology* 114, no. 4 (2001): 493–510.

Blom, Jan-Dirk. *Hallucinations: Research and Practice.* New York [etc.]: Springer, 2012.

Blonk, Jaap, and René van Peer. "Sounding the Outer Limits." *Leonardo Music Journal* 15 (2005): 62–68.

Boethius, Anicius Manlius Severinus. *Fundamentals of Music.* Edited by Claude V. Palisca, translated by Calvin M. Bower. New Haven: Yale University Press, 1989.

Bregman, Albert S. *Auditory Scene Analysis: The Perceptual Organization of Sound.* Cambridge, MA: MIT Press, 1990.

Bregman, A. S., and A. I. Rudnicky. "Auditory Segregation: Stream or Streams?" *Journal of Experimental Psychology. Human Perception and Performance* 1, no. 3 (1975): 263–267.

Breton, André, and André Parinaud. *Conversations: The Autobiography of Surrealism.* New York: Paragon House, 1993.

Breton, André, Richard Seaver, and Helen R. Lane. *Manifestoes of Surrealism.* Ann Arbor: University of Michigan Press, 2008.

Brown, Steven, and Ulrik Volgsten. *Music and Manipulation: On the Social Uses and Social Control of Music.* New York; Oxford: Berghahn Books, 2006.

Brugger, Peter. "From Haunted Brain to Haunted Science: A Cognitive Neuroscience View of Paranormal and Pseudoscientific Thought." In *Hauntings and Poltergeists: Multidisciplinary Perspectives,* edited by J. Houran and R. Lange, 195–213. Berkeley: Continuum, 2004.

Bücher, Karl. *Arbeit und Rhythmus.* Leipzig: Hirzel, 1896.

Burnham, Jack. "Systems Esthetics." *Artforum* 1 (1968): 30–35.

Burroughs, William S. *The Adding Machine: Selected Essays.* New York: Grove Press, 2013.

Burt, Ramsay. *The Judson Dance Theatre Performative Traces.* London: Routledge, 2011.

Cambouropoulos, Emilios, and Costas Tsougras. "Auditory Streams in Ligeti's Continuum: A Theoretical and Perceptual Approach." *Journal of Interdisclipinary Music Studies* 3, nos 1 and 2 (2009): 119–137.

Cardeña, Etzel, and Jane Beard. "Truthful Trickery: Shamanism, Acting and Reality." *Performance Research Performance Research* 1, no. 3 (1996): 31–39.
Carpenter, Siri. "ESP Findings Send Controversial Message." *Science News* 156, no. 5 (1999): 70.
Carroll, Peter J. *Apophenion: A Chaos Magic Paradigm*. London: Lightning Source, 2008.
Casa, Eric La, and Cédric Peyronnet. "Zones Portuaires." Malaysia: Herbal International– Concrete Disc 1304-2, 2013.
Castillo, Richard J. "Culture, Trance, and the Mind-Brain." *Anthropology of Consciousness* 6, no. 1 (1995): 17–34.
Changizi, Mark A. *Harnessed: How Language and Music Mimicked Nature and Transformed Ape to Man*. Dallas: BenBella Books, 2011.
Chave, Anna C. "Revaluing Minimalism: Patronage, Aura, and Place." *Art Bulletin* 90, no. 3 (2008): 466–486.
Cherry, E. Colin. *On Human Communication*. Cambridge, MA: MIT Press, 1966.
Cherry, E. Colin, Morris Halle, and Roman Jakobson. "Toward the Logical Description of Languages in Their Phonemic Aspect." *Language* 29, no. 1 (1953): 34–46.
Clark, Andy. "Whatever Next? Predictive Brains, Situated Agents, and the Future of Cognitive Science." *Behavioral and Brain Sciences* 36, no. 3 (2013): 181–204. doi:10.1017/S0140525X12000477.
Clark, Andy. "Perception as Controlled Hallucination | A Conversation with Andy Clark." Interview - Podcast. *EdgeCast*. EdgeCast (Edge.org), 2019. https://www.edge.org/conversation/andy_clark-perception-as-controlled-hallucination. Accessed 27 October 2019.
Clark, Andy. *Surfing Uncertainty: Prediction, Action, and the Embodied Mind*. Oxford: Oxford University Press, 2019.
Clarke, Eric F. "Rhythm and Timing in Music." In *The Psychology of Music*, edited by Diana Deutsch, 473–500. San Diego: Academic Press, 1999.
Clayton, Martin. "What Is Entrainment? Definition and Applications in Musical Research." *Empirical Musicology Review* 7, nos 1–2 (2012): 49–56.
Clayton, Martin, Rebecca Sager, and Udo Will. "In Time with the Music: The Concept of Entrainment and Its Significance for Ethnomusicology." *European Meetings in Ethnomusicology* 11 (2005): 1–82.
Collins, Nicholas, and Tom Plsek. "Nicholas Collins Responds to Reviewer Tom Plsek." *Computer Music Journal* 11, no. 1 (1987): 14–16. doi:10.2307/3680171.
Connolly, Brian. "The Inner Ear as a Musical Instrument." *The Journal of the Acoustical Society of America* 138, no. 3 (2015): 1935.
Conrad, Klaus. *Die beginnende schizophrenie: versuch einer gestaltanalyse des wahns*. Stuttgart: Georg Thieme, 1958.
Cornwell, Regina. "Paul Sharits: Illusion and Object." *ArtForum* 10, no. 1 (1971): 61.
Coughlan, Aidan. *The Broadsheet Book of Unspecified Things That Look like Ireland*. New Island Books, 2013.
Cytowic, Richard E. *Synesthesia—A Union of the Senses*. 2nd edn. Cambridge, MA: MIT Press, 2002.
Dahlhaus, Carl. *The Idea of Absolute Music*. Chicago; London: University of Chicago Press, 1989.
Dalglish, William E. "The Hocket in Medieval Polyphony." *The Musical Quarterly* 55, no. 3 (1969): 344–363.

Dalglish, William. "The Origin of the Hocket." *Journal of the American Musicological Society* 31, no. 1 (1978): 3–20.
Daly, Scott. "The Ganzfeld as a Canvas for Neurophysiologically Based Artworks." *Leonardo* 17, no. 3 (1984): 172–175.
Davis, Mary E. *Erik Satie*. London: Reaktion Books, 2007.
Deleuze, Gilles. *Difference and Repetition*. London [etc.]: Continuum, 2004.
Denham, A. E. "The Moving Mirrors of Music: Roger Scruton Resonates with Tradition." *Music & Letters* 80, no. 3 (1999): 411–432.
Denham, Susan L., and István Winkler. "Predictive Coding in Auditory Perception: Challenges and Unresolved Questions." *European Journal of Neuroscience*, December 18, 2017. doi:10.1111/ejn.13802.
Denham, S. L., and I. Winkler. "The Role of Predictive Models in the Formation of Auditory Streams." *Journal of Physiology, Paris* 100, nos 1–3 (2006): 154–170.
Deutsch, Diana. *Musical Illusions and Paradoxes*. La Jolla, CA: Philomel, 1995.
Deutsch, Diana. "Grouping Mechanisms in Music." In *The Psychology of Music*, 299–348, 1999.
Deutsch, Diana. "Phantom Words and Other Curiosities." La Jolla, CA: Philomel Records, 2003.
Deutsch, Diana. "Phantom Words | Psychology Today." *Psychology Today*, 2009. https://www.psychologytoday.com/us/blog/illusions-and-curiosities/200906/phantom-words. Accessed October 30, 2019.
Deutsch, Diana. *The Psychology of Music*. London: Academic Press, 2013.
Deutsch, Diana. "The Tritone Paradox: An Influence of Language on Music Perception." *Music Perception: An Interdisciplinary Journal* 8, no. 4 (1991): 335–347.
DeWeese, Michael R., and Anthony M. Zador. "Neural Gallops Across Auditory Streams." *Neuron* 48, no. 1 (2005): 5–7.
Dissanayake, Ellen. "Art as a Human Behavior: Toward an Ethological View of Art." *Journal of Aesthetics and Art Criticism* 38, no. 4 (1980): 397–406.
Dissanayake, Ellen. *Homo Aestheticus: Where Art Comes From and Why*. New York: Free Press, 1992.
Dixon, D. "I Hear Dead People: Science, Technology and a Resonant Universe." *Social and Cultural Geography* 8, no. 5 (2007): 719–733.
Edelman, Gerald M., and Giulio Tononi. *A Universe of Consciousness: How Matter Becomes Imagination*. New York: Basic Books, 2000.
Eliade, Mircea. *Shamanism: Archaic Techniques of Ecstasy, Etc*. London: Routledge & Kegan Paul, 1964.
Emmerson, S. T. "From Dance! To 'Dance': Distance and Digits." *Computer Music Journal* 25 (2001): 13–20.
Erickson, Robert. *Sound Structure in Music*. Berkeley: University of California Press, 1975.
Erlmann, Veit. "Trance and Music in the Hausa Bòorii Spirit Possession Cult in Niger." *Ethnomusicology* 26, no. 1 (1982): 49–58.
Farina, Almo, Emanuele Lattanzi, Rachele Malavasi, Nadia Pieretti, and Luigi Piccioli. "Avian Soundscapes and Cognitive Landscapes: Theory, Application and Ecological Perspectives." *Landscape Ecology* 26, no. 9 (2011): 1257–1267.
Feld, Steven. *Sound and Sentiment: Birds, Weeping, Poetics, and Song in Kaluli Expression*. Philadelphia: University of Pennsylvania Press, 1989.

Feld, Steven, and Institute of Papua New Guinea Studies. *Music of the Kaluli*. [Boroko, Papua New Guinea]: Institute of Papua New Guinea Studies, 1981.

Fernandes, Frederico, and Enzo Minarelli. *Polypoetry 30 Years 1987–2017*. Londrina: Edeul, 2018.

Fields, M. C., L. V. Marcuse, J.-Y. Yoo, and S. Ghatan. "Palinacousis, Palinacousis: Seven New Cases." *Journal of Clinical Neurophysiology* 35, no. 2 (2018): 173–176.

Ford, Simon. *Wreckers of Civilisation: The Story of COUM Transmissions & Throbbing Gristle*. London: Black Dog, 1999.

Fries, Jan. *Visual-Magick: A Manual of Freestyle Shamanism*. Oxford: Mandrake, 1992.

Friston, Karl. "The Free-Energy Principle: A Unified Brain Theory?" *Nature Reviews. Neuroscience* 11, no. 2 (2010): 127.

Friston, Karl J., Klaas Enno Stephan, Read Montague, and Raymond J. Dolan. "Computational Psychiatry: The Brain as a Phantastic Organ." *The Lancet. Psychiatry* 1, no. 2 (July 2014): 148–158.

Frith, Christopher. *How the Brain Creates Our Mental World*. Chicester: Wiley-Blackwell, 2007.

Frosch, F. G. "Hum and Otoacoustic Emissions May Arise Out of the Same Mechanisms." *Journal of Scientific Exploration* 27, no. 4 (2013): 603–624.

Gamboni, Dario. *Potential Images: Ambiguity and Indeterminacy in Modern Art*. London: Reaktion, 2004.

Gamboni, Dario. "Stumbling Over and Upon Art." *Cabinet: A Quarterly of Art and Culture* 19 (Fall 2005): 58–61. http://www.cabinetmagazine.org/issues/19/gamboni.php.

Geiersbach, Frederick J. "Making the Most of Minimalism in Music." *Music Educators Journal* 85, no. 3 (1998): 26–49.

Gendreau, Michael. *Parataxes: fragments pour une architecture des espaces sonores*. Paris; Lausanne: Editions van Dieren (collection Rip On/Off), 2010.

Gibson, William. *Pattern Recognition*. New York: G.P. Putnam's Sons, 2003.

Gioia, Ted. *Work Songs*. Durham, NC: Duke University Press, 2006.

Gogala, Matija, and K. Riede. "Time Sharing of Song Activity by Cicadas in Temengor Forest Reserve, Hulu Perak, and in Sabah, Malaysia." *Malayan Nature Journal* 34 (1995): 3–4.

Gombrich, E. H. *Art and Illusion: A Study in the Psychology of Pictorial Representation*. London: Phaidon, 2002.

Gombrowicz, Witold. *Diary, Volume 1*. Evanston, IL: Northwestern University Press, 1988.

Grant, Morag Josephine. "Experimental Music Semiotics." *International Review of the Aesthetics and Sociology of Music* 34, no. 2 (2003): 173–191.

Grauer, Victor A. *Some Song-Style Clusters—a Preliminary Study*. Society for Ethnomusicology, 1965.

Grauer, Victor A. *Sounding the Depths: Tradition and the Voices of History*. CreateSpace, 2011.

Gray, John. *The Immortalization Commission: Science and the Strange Quest to Cheat Death*. New York: Farrar, Straus and Giroux, 2011.

Greenfield, M. D., and J. Schul. "Mechanisms and Evolution of Synchronous Chorusing: Emergent Properties and Adaptive Functions in Neoconocephalus Katydids (Orthoptera: Tettigoniidae)." *Journal of Comparative Psychology* 122, no. 3 (2008): 289–297.

Gregory, R. L. *The Intelligent Eye*. London: Weidenfeld & Nicolson, 1970.

Gregory, R L. *Visual Perception*. London: Oxford University Press, 1974.

Griffin, Juliet D., and Paul C. Fletcher. "Predictive Processing, Source Monitoring, and Psychosis." *Annual Review of Clinical Psychology* 13 (2017): 265–289.

Grush, R. "The Emulation Theory of Representation: Motor Control, Imagery, and Perception." *Behavioral and Brain Sciences* 27, no. 3 (2004): 377–396.

Guthrie, Stewart Elliot. *Faces in the Clouds: A New Theory of Religion.* New York: Oxford University Press, 1995.

Gwiazda, Henry. "Sound Is Not Enough." *Leonardo Music Journal* 22 (2012): 57–58.

Handel, Stephen. *Listening: An Introduction to the Perception of Auditory Events.* Cambridge, MA: MIT Press, 1989.

Haraway, Donna Jeanne. *The Companion Species Manifesto: Dogs, People, and Significant Otherness.* Chicago: Prickly Paradigm, 2003.

Harley, James. "Book Review: Persepolis." *Computer Music Journal* 25, no. 1 (2001): 92–93.

Hegarty, Paul. *Noise/Music: A History.* New York: Continuum, 2007.

Heise, G. A., and George A. Miller. "An Experimental Study of Auditory Patterns." *The American Journal of Psychology* 64, no. 1 (1951): 68–77.

Helmholtz, Hermann. *On the Sensations of Tone.* Translated by Alexander John Ellis. New York: Dover, 1954.

Heron, Woodburn. "The Pathology of Boredom." *Scientific American* 196, no. 1 (1957): 52–57.

Hilgard, Ernest R. "Divided Consciousness and Dissociation." *Consciousness and Cognition* 1, no. 1 (1992): 16–31.

Hine, Phil. "The Magical Use of the Voice." *Chaos International.* https://www.scribd.com/document/34442903/Magical-Use-of-Voice-Phil-Hine. Accessed October 26, 2019.

Hine, Phil, and Peter J. Carroll. *Condensed Chaos: An Introduction to Chaos Magic.* Tempe, AZ: Original Falcon Press, 2010.

Hohwy, Jakob. *The Predictive Mind.* Oxford: Oxford University Press, 2013.

Hooper, Judith, and Dick Teresi. *Would the Buddha Wear a Walkman?: A Catalogue of Revolutionary Tools for Higher Consciousness.* New York: Simon & Schuster, 1990.

Houran, James, and Rense Lange. *Hauntings and Poltergeists: Multidisciplinary Perspectives.* Jefferson, NC: McFarland, 2001.

Howard, David M., and J. A. S. Angus. *Acoustics and Psychacoustics.* Oxford: Focal, 1996.

Huizinga, Johan. *Homo Ludens: A Study of the Play-Element in Culture.* Boston: Beacon Press, 1955.

Hung, Wu, Guangzhou Triennial, Wang Huangsheng, and Feng Boyi. *Reinterpretation: A Decade of Experimental Chinese Art (1990–2000).* Chicago; Guangzhou: Art Media Resources; Guangdong Museum of Art, 2002.

Hyde, Lewis. *Trickster Makes This World: Mischief, Myth, and Art.* New York: Farrar, Straus and Giroux, 1998.

Ibn al-Haytham. *The Optics of Ibn Al-Haytham. Books I–III: On Direct Vision.* Translated by A. I. Sabra. London: Warburg Institute, University of London, 1989.

Jakobovits, Leon. A. "Effects of Repeated Stimulation on Cognitive Aspects of Behavior: Some Experiments on the Phenomenon of Semantic Satiation." Thesis, Department of Psychology, McGill University, Montreal, April 1962. https://central.bac-lac.gc.ca/.item?id=TC-QMM-113683&op=pdf&app=Library. Accessed January 3, 2020.

Jones, Arthur Morris. *Studies in African Music.* London: Oxford University Press, 1961.

Jones, Leslie, Isabelle Dervaux, and Susan Laxton. *Drawing Surrealism.* Los Angeles County Museum of Art, 2012.

Jordania, Joseph. *Why Do People Sing?: Music in Human Evolution.* Tbilisi [Georgia]; Logos: Tbilisi Ivane Javakhishvili State University, 2011.

Joseph, Branden Wayne. *Beyond the Dream Syndicate: Tony Conrad and the Arts after Cage: (A "Minor History")*. New York: Zone Books, 2011.
Kaizen, William. "Steps to an Ecology of Communication: 'Radical Software', Dan Graham, and the Legacy of Gregory Bateson." *Art Journal/College Art Association of America* 67, no. 3 (2008): 86–107.
Kandel, Eric R. *The Age of Insight: The Quest to Understand the Unconscious in Art, Mind, and Brain: From Vienna 1900 to the Present*. New York: Random House, 2012.
Kane, Brian. *Sound Unseen: Acousmatic Sound in Theory and Practice*. Oxford University Press, 2016.
Kanungo, Rabindranath, and W. E. Lambert. "Semantic Satiation and Meaningfulness." *The American Journal of Psychology* 76, no. 3 (1963): 421–428.
Karl, Friston. "A Free Energy Principle for Biological Systems." *Entropy Entropy* 14, no. 11 (2012): 2100–2121.
Kato, Masaharu, and Ryoko Mugitani. "Pareidolia in Infants." *PLOS ONE* 2 (2015): 1–9.
Kippen, James, and Bernard Bel. "Computers, Composition, and the Challenge of 'New Music' in Modern India." *Leonardo Music Journal* (1994): 79–84.
Kirk, Jonathon. "Otoacoustic Emissions as a Compositional Tool." *International Computer Music Conference Proceedings* (2010), 316–318.
Kirkegaard, Jacob, and Museet for Samtidskunst (Roskilde). *Earside Out*. Roskilde: Museet for Samtidskunst, 2015.
Kittler, Friedrich A. *Gramophone, Film, Typewriter*. Stanford, CA: Stanford University Press, 1999.
Koelsch, Stefan, Peter Vuust, and Karl Friston. "Predictive Processes and the Peculiar Case of Music." *Trends in Cognitive Sciences* 23, no. 1 (January 1, 2019): 63–77.
Krause, Bernard L. *The Great Animal Orchestra: Finding the Origins of Music in the World's Wild Places*. London: Profile Books, 2013.
Krause, Bernie. "The Niche Hypothesis: A Virtual Symphony of Animal Sounds, the Origins of Musical Expression and the Health of Habitats." *Soundscape Newsletter (World Forum for Acoustic Ecology)* (June 6, 1993), 6–10.
Krauss, R. "Paul Sharits: Stop Time." *ArtForum* 11, no. 8 (1973): 60–61.
Kubik, Gerhard. "The Phenomenon of Inherent Rhythms in East and Central African Instrumental Music." *African Music* 3 (1962): 33–42.
Kubik, Gerhard. *Theory of African Music, Volume I*. Chicago: University of Chicago Press, 2010.
Kubik, Gerhard. *Theory of African Music, Volume II*. Chicago; London: University of Chicago Press, 2010.
Labelle, Brandon, and Sabine Brietsameter. "Questionnaire." *Soundscape* 5, no. 2 (2002): 11–14.
Lander, L. *Beyond the Dial: An Investigation into the Electronic Voice Phenomenon and Other Esoteric Technological Transceivings*. Gwynedd, UK: Orb Editions, 2009.
Lewis-Williams, J. David. *The Mind in the Cave: Consciousness and the Origins of Art*. London: Thames & Hudson, 2002.
Lewis-Williams, J. David, and D. G. Pearce. *Inside the Neolithic Mind: Consciousness, Cosmos and the Realm of the Gods*. London: Thames & Hudson, 2009.
López, Francisco. "Against the Stage." *Zehar Magazine* 53 (2004): 36–39.
López, Francisco. "Environmental Sound Matter." In *Audio Culture: Readings in Modern Music*. Edited by Christoph Cox and Daniel Warner, 82–87. New York: Continuum, 2004.

López, Francisco. "Music Dematerialized?" *Journal of Sonic Studies* 7 (2014).
López, Francisco. "Sonic Creatures." http://www.franciscolopez.net/pdf/creatures.pdf 2019.
Loubet, Emmanuelle, and Marc Couroux. "Laptop Performers, Compact Disc Designers, and No-Beat Techno Artists in Japan: Music from Nowhere." *Computer Music Journal* 24, no. 4 (2000): 19–32.
Lucier, Alvin and Robert Ashley. *Music 109: Notes on Experimental Music*. Middletown, OH: Wesleyan University Press, 2012.
Lupyan, Gary. "Cognitive Penetrability of Perception in the Age of Prediction: Predictive Systems are Penetrable Systems." *Review of Philosophy and Psychology* 6, no. 4 (2015): 547–569.
MacDonald, John, and Harry McGurk. "Visual Influences on Speech Perception Processes." *Perception & Psychophysics* 24, no. 3 (October 1978): 253–257.
Mace, Dean T. "Marin Mersenne on Language and Music." *Journal of Music Theory* 14, no. 1 (1970): 2.
Mace, Stephen. *Stealing the Fire from Heaven: A Technique for Creating Individual Systems of Sorcery*. Phoenix, AZ: Dagon Productions, 2003.
Macknik, Stephen L., S. Martinez-Conde, and Sandra Blakeslee. *Sleights of Mind: What the Neuroscience of Magic Reveals about Our Everyday Deceptions*. New York: Picador, 2011.
Maclagan, David. *Line Let Loose: Scribbling, Doodling and Automatic Drawing*. London: Reaktion Books, 2014.
Malavasi, Rachele, and Almo Farina. "Neighbours' Talk: Interspecific Choruses among Songbirds." *Bioacoustics* 22, no. 1 (2013): 33–48.
Marsching, Jane D. "Orbs, Blobs, and Glows: Astronauts, UFOs, and Photography." *Art Journal* 62, no. 3 (2003): 56–65.
Massumi, Brian. *Parables for the Virtual: Movement, Affect, Sensation*. Durham, NC; London: Duke University Press, 2002.
Mavromatis, Andreas. *Hypnagogia: The Unique State of Consciousness Between Wakefulness and Sleep*. London: Thyrsos Press, 2010.
McAdams, Stephen, and Albert Bregman. "Hearing Musical Streams." *Computer Music Journal* 3, no. 4 (1979): 26–60.
McNeill, William H. *Keeping Together in Time: Dance and Drill in Human History*. Cambridge, MA: Harvard University Press, 1997.
McNulty, S. J., and P. Williams. "If You're Going to Vegas, Watch Out for the Texas Sharp Shooter." *BMJ (Online)* 345 (2012).
Merckelbach, Harald, and Vincent van de Ven. "Another White Christmas: Fantasy Proneness and Reports of Hallucinatory Experiences in Undergraduate Students." *Journal of Behavior Therapy and Experimental Psychiatry* 32, no. 3 (2001): 137–144.
Mersenne, Marin. *Harmonie Universelle, contenant la Théorie et la Pratique de la Musique*. Paris: Harmonie Universelle. First edition online from Gallica (Paris, 1636–1637). Translation to English by Roger E. Chapman (The Hague, 1957). https://gallica.bnf.fr/ark:/12148/bpt6k5471093v. Accessed January 3, 2020.
Meschiari, Matteo. "Roots of the Savage Mind. Apophenia and Imagination as Cognitive Process." *Quaderni di Semantica* 2, no. 9 (2009): 1–39.
Millar, Jonathan. "Conceptual Writing." *Art Monthly* 361 (2012): 10–13.
Miller, G. A., and J. C. R. Licklider. "The Intelligibility of Interrupted Speech." *Journal of the Acoustical Society of America* 22 (1950): 167–173. doi:10.1121/1.1906584.
Miller, George A. "The Trill Threshold." *Physics Today* 3, no. 10 (1950): 32.

Miller, George A., and George A. Heise. "The Trill Threshold." *The Journal of the Acoustical Society of America* 22, no. 5 (1950): 637–638.

Milner, Peter M. "The Mind and Donald O. Hebb." *Scientific American* 268, no. 1 (1993): 124–129.

Moles, Abraham A. *Information Theory and Esthetic Perception*. Urbana: University of Illinois Press, 1966.

MoMA. "The Collection | MoMA." *MoMa.Org/Collection*. https://www.moma.org/ Accessed October 26, 2019.

Moore, Christopher. "Apophenia—May 25, 2012," 2012. https://www.christophergmoore.com/post/apophenia. Accessed October 26, 2019.

Myers, F. W. H., and Susy Smith. *Human Personality and its Survival of Bodily Death*. New York: Dover Publications, 2005.

Nattiez, Jean-Jacques. "Some Aspects of Inuit Vocal Games." *Ethnomusicology* 27, no. 3 (1983): 457–475.

Nattiez, Jean-Jacques. "The Rekkukara of the Ainu (Japan) and the Katajjaq of the Inuit (Canada): A Comparison." *The World of Music* 25, no. 2 (1983): 33–44.

Nattiez, Jean-Jacques. "Inuit Throat-Games and Siberian Throat Singing: A Comparative, Historical, and Semiological Approach." *Ethnomusicology* 43, no. 3 (1999): 399–418.

Nechvatal, Joseph. *Towards an Immersive Intelligence: Essays on the Work of Art in the Age of Computer Technology and Virtual Reality; 1993–2006*. New York: Edgewise, 2009.

Neher, Andrew. "A Physiological Explanation of Unusual Behaviour in Ceremonies Involving Drums." *Human Biology* 34, no. 2 (1962): 151–160.

Neher, Andrew. "Auditory Driving Observed with Scalp Electrodes in Normal Subjects." *Electroencephalography and Clinical Neurophysiology* 13, no. 3 (1961): 449–451.

Nin, Anais. *Seduction of the Minotaur*. Denver: Swallow, 1961.

Nixon, Mignon. "Dream Dust." *October* 116 (2006): 63–86.

Nketia, J. H. Kwabena. "The Hocket-Technique in African Music." *Journal of the International Folk Music Council* 14 (1962): 44–52.

Núñez, Nicolás, Deborah Middleton, and Ronan J. Fitzsimons. *Anthropocosmic Theatre: Rite in the Dynamics of Theatre*. London: Taylor & Francis, 2005.

O'Callaghan, Casey. *Sounds: A Philosophical Theory*. Oxford: Oxford University Press, 2007.

Ondobaka, Sasha, and Harold Bekkering. "Hierarchy of Idea-Guided Action and Perception-Guided Movement." *Frontiers in Psychology* 3 (December 27, 2012): 579.

Palombini, Carlos. "Machine Songs V: Pierre Schaeffer: From Research into Noises to Experimental Music." *Computer Music Journal* 17, no. 3 (1993): 14–19.

Palombini, Carlos. "Pierre Schaeffer, 1953: Towards an Experimental Music." *Music and Letters* 74, no. 4 (November 1, 1993): 542–557. doi:10.1093/ml/74.4.542.

Pandey, Ashish. *Encyclopaedic Dictionary of Music*. Delhi: Isha Books, 2005.

Panzner, Joe. *The Process That is the World: Cage/Deleuze/Events/Performances*. New York: Bloomsbury Academic, 2017.

Pater, Walter. *The Renaissance: Studies in Art and Poetry; the 1893 Text*. Berkeley: University of California Press, 1980.

Paton, Bryan, Josh Skewes, Chris Frith, and Jakob Hohwy. "Skull-Bound Perception and Precision Optimization Through Culture." *Behavioral and Brain Sciences* 36, no. 3 (2013): 222. doi:10.1017/S0140525X12002191.

Pec, Ondrej, Bob Petr, and Jiri Raboch. "Dissociation in Schizophrenia and Borderline Personality Disorder." *Neuropsychiatric Disease and Treatment* 10 (2014): 487–491.

Phillips-Silver, Jessica, C. Athena Aktipis, and Gregory A. Bryant. "The Ecology of Entrainment: Foundations of Coordinated Rhythmic Movement." *Music Perception* 28, no. 1 (2010): 3–14.

Pick, Herbert L., and Richard D. Walk. *Intersensory: Perception and Sensory Integration*. New York [etc.]: Plenum, 1981.

Plantenga, Bart. *Yodel-Ay-Ee-Oooo The Secret History of Yodeling Around the World*. Routledge, 2013.

Plantenga, Bart. *Yodel in Hi-Fi: From Kitsch Folk to Contemporary Electronica*. Madison, WI: University of Wisconsin Press, 2013.

Platz, Robert H. P., and Frances Wharton. "More Than Just Notes: Psychoacoustics and Composition." *Leonardo Music Journal* 5 (1995): 23–28.

Poe, Edgar Allan, Stuart Levine, and Susan Levine. *The Short Fiction of Edgar Allan Poe: An Annotated Edition*. Urbana; Chicago: University of Illinois Press, 1990.

Poss, R. "Distortion is Truth." *Leonardo Music Journal* 8 (1998): 45–48.

Pressnitzer, D., C. Suied, and S. A. Shamma. "Auditory Scene Analysis: The Sweet Music of Ambiguity." *Frontiers in Human Neuroscience* 5 (2011): Article 158.

Probst, R., B. L. Lonsbury-Martin, and G. K. Martin. "A Review of Otoacoustic Emissions." *The Journal of the Acoustical Society of America* 89, no. 5 (1991): 2027–2067.

Psychogeographical Dip CD | Discogs. US: GD Stereo - GD 013, 1997.

Radcliffe-Brown, Alfred Reginald. *The Andaman Islanders*. New York: Free Press of Glencoe, 1964.

Ramachandran, V. S., and W. Hirstein. "The Science of Art: A Neurological Theory of Aesthetic Experience." *Journal of Consciousness Studies* 7, no. 6/7 (1999): 15–51.

Read, Herbert Edward. *To Hell with Culture, and Other Essays on Art and Society*. London: Routledge & Kegan Paul, 2002.

Régénération—Dégénérescence. Kaon. France: Kaon, 1997.

Reich, Steve, and Paul Hillier. *Writings on Music, 1965–2000*. New York; Oxford: Oxford University Press, 2004.

Robindoré, Brigitte, and Curtis Roads. "Forays into Uncharted Territories: An Interview with Curtis Roads." *Computer Music Journal* 29, no. 1 (2005): 11–20.

Rondeau, James E. "Clown Torture, 1987 by Bruce Nauman." *Art Institute of Chicago Museum Studies* 25, no. 1 (1999): 62.

Rorschach, Hermann. *Psychodiagnostik; Methodik und Ergebnisse eines wahrnehmungsdiagnostischen Experiments; Deutenlassen von Zufallsformen*. Bern: Bircher, 1921.

Rosenboom, David. "Propositional Music: On Emergent Properties in Morphogenesis and the Evolution of Music. Part I: Essays, Propositions and Commentaries." *Leonardo* 30, no. 4 (1997): 291–297.

Rosenfeld, Edward. *The Book of Highs; 250 Ways to Alter Consciousness Without Drugs*. New York: Quadrangle, 1973.

Rothenberg, David. *Bug Music: How Insects Gave Us Rhythm and Noise*. New York: St Martin's Press, 2013.

Rothenberg, David. *Why Birds Sing: A Journey into the Mystery of Bird Song*. New York: Basic Books, 2005.

Ruppé, L., G. Clément, A. Herrel, L. Ballesta, T. Décamps, L. Kéver, and E. Parmentier. "Environmental Constraints Drive the Partitioning of the Soundscape in Fishes."

Proceedings of the National Academy of Sciences of the United States of America 112, no. 19 (2015): 6092–6097.

Sacks, Oliver. *Musicophilia: Tales of Music and the Brain*. London: Picador, 2008.

Sagan, Carl. *The Demon-Haunted World*. New York: Random House, 1997.

Schaeffer, Pierre. *Treatise on Musical Objects: An Essay Across Disciplines*. Translated by Christine North and John Dack. Oakland, CA: University of California Press, 2017.

Schafer, R. Murray. *The Tuning of the World*. Toronto: McClelland and Stewart, 1977.

Schäfer, Helmut. "Kraków, 05. 03.05| AudioTong." *Audio Tong, Bandcamp*, 2007. https://audiotong.bandcamp.com/album/krak-w-050305. Accessed December 26, 2018.

Schapiro, M. "The Liberating Quality of the Avant-Garde." *Art News* 4, no. 56 (1957): 36–42.

Scharine, Angelique, and Tomasz Letowski. "Auditory Conflicts and Illusions."*Helmet-Mounted Displays: Sensation, Perception and Cognition Issues*. Edited by C. E. Rash, M. B. Russo, R. R. Letowski and E. T. Schmeisser, 579–598. US Army Aeromedical Research Laboratory, 2009. doi:10.13140/2.1.3684.4804.

Scherzinger, Martin. "György Ligeti and the Aka Pygmies Project." *Contemporary Music Review* 25, no. 3 (2006): 227–262.

Schreber, Daniel Paul. *Memoirs of my Nervous Illness*. New York: New York Review Books, 2000.

Seth, Anil K. "Anil Seth: Your Brain Hallucinates Your Conscious Reality." *TED Talk*, 2018. https://www.ted.com/talks/anil_seth_how_your_brain_hallucinates_your_conscious_reality. Accessed October 21, 2019.

Seth, Anil K. "The Neuroscience of Reality." *Scientific American* 321, no. 3 (September 2019): 1–11.

Seth, Anil K. "From Unconscious Inference to the Beholder's Share: Predictive Perception and Human Experience." *European Review* 27, no. 3 (2019): 378–410.

Sharits, P. "General Statement for 4th International Experimental Film Festival, Knokke-Le Zoute." *Film Culture* 47 (1969): 13.

Shklovskii, Viktor, Lee T. Lemon, Marion J. Reis, Gary Saul Morson, Viktor Shklovskii, Viktor Shklovskii, B. V. Tomashevskii, and B. Eikhenbaum. *Russian Formalist Criticism: Four Essays*. Lincoln: University of Nebraska Press, 2012.

Siebzehn Bis ∞ (2015, File) Discogs. *Crónica – Crónica 096~2015*. Portugal: Crónica – Crónica 096~2015, 2015.

Smith, Daniel B. *Muses, Madmen and Prophets: Rethinking the History, Science, and Meaning of Auditory Hallucination*. New York: The Penguin Press, 2007.

Smith, William S. "A Concrete Experience of Nothing: Paul Sharit's Flicker Films." *Res/Peabody Museum of Archaeology and Ethnology and the Harvard University Art Museums* 55/56 (2009): 279–293.

Snyder, J. S., M. K. Gregg, D. M. Weintraub, and C. Alain. "Attention, Awareness, and the Perception of Auditory Scenes." *Frontiers in Psychology* 3 (2012): 1–15.

Stepputat, Kendra, and Rudi Samapati. "Performing Kecak: A Balinese Dance Tradition between Daily Routine and Creative Art." *Yearbook for Traditional Music* 44 (2012): 49–70.

Stitt, A., M. Vason, R. Hunter, L. Keidan, and R. Athey. *Small Time Life*. London: Black Dog, 2001.

Stowell, Dan, and Mark Plumbley. *Characteristics of the Beatboxing Vocal Style*. Technical report, Queen Mary University of London, 2008.

Strogatz, Steven Henry. *Sync: The Emerging Science of Spontaneous Order*. New York: Hyperion, 2003.
Suzuki, Dean. "A Polypoetical Collision: Text-Sound Composition, Minimal and Postminimal Music, Rock and Jazz." In *Polypoetry 30 Years 1987-2017*. Edited by E. Minarelli and F. Fernandes, 391. Londrina: Edeul, 2018.
Swanson, L. R. "The Predictive Processing Paradigm has Roots in Kant." *Frontiers in Systems Neuroscience* 10 (2016). doi:10.3389/fnsys.2016.00079.
Taine, Hippolyte, and T. D. Haye. *On Intelligence*. New York: Holt & Williams, 1872.
Thomassen, Sabine, and Alexandra Bendixen. "Subjective Perceptual Organization of a Complex Auditory Scene." *The Journal of the Acoustical Society of America* 141, no. 1 (January 2017): 265–276. doi:10.1121/1.4973806.
Thompson, Tok. "Beatboxing, Mashups, and Cyborg Identity: Folk Music for the Twenty-First Century." *Western Folklore* 70, no. 2 (2011): 171–193.
Tone, Yasunao. *Yasunao Tone: Noise Media Language*. [Los Angeles, CA]: Errant Bodies Press, 2007.
Truax, Barry. "Composition and Diffusion: Space in Sound in Space." *Organised Sound* 3, no. 2 (1998): 141–146.
Turner, Charles. "Xenakis in America." Thesis, City University of New York, 2014.
Turner, Victor W. *The Ritual Process: Structure and Anti-Structure*. Chicago: Aldine, 1969.
Uexküll, Jakob von. *A Foray into the Worlds of Animals and Humans; with, A Theory of Meaning*. Minneapolis: University of Minnesota Press, 2010.
Vaughn, Kathryn. "Exploring Emotion in Sub-Structural Aspects of Karelian Lament: Application of Time Series Analysis to Digitized Melody." *Yearbook for Traditional Music* 22 (1990): 106–122.
Velitchkina, Olga. "The Role of Movement in Russian Panpipe Playing." *Ethnomusicology OnLine* 2 (1996).
da Vinci, Leonardo. *"Trattato Della Pittura" Di Leonardo Da Vinci*. Edited by Angelo Borzelli. Lanciano: G. Carabba, 1947.
Vitebsky, Piers. *Secrets of the Shaman*. Alexandria; London: Time-Life Books; Duncan Baird, 1995.
Vitebsky, Piers. *The Shaman: Voices of the Soul Trance, Ecstasy, and the Healing from Siberia to the Amazon*. London: Duncan Baird, 1995.
Walker, T. J. "Acoustic Synchrony: Two Mechanisms in the Snowy Tree Cricket." *Science* 166, no. 3907 (1969): 891–894.
Wallin, Nils Lennart, Bjorn Merker, and Steven Brown. *The Origins of Music*. Cambridge; London: MIT Press, 2005.
Wallisch, Pascal. "Here's Why People Saw 'the Dress' Differently." *Slate*. https://slate.com/technology/2017/04/heres-why-people-saw-the-dress-differently.html. Accessed October 31, 2019.
Walter, V. J., and W. Grey Walter. "The Central Effects of Rhythmic Sensory Stimulation." *Electroencephalography and Clinical Neurophysiology* 1, no. 1–4 (1949): 57–86.
Warren, R. M. "Illusory Changes of Distinct Speech upon Repetition—the Verbal Transformation Effect." *British Journal of Psychology* 52 (1961): 249–258.
Wegner, Ulrich. "Cognitive Aspects of Amadinda Xylophone Music from Buganda: Inherent Patterns Reconsidered." *Ethnomusicology* 37, no. 2 (1993): 201–241.
Weiss, Allen S. *Varieties of Audio Mimesis: Musical Evocations of Landscape*. [Los Angeles]: Errant Bodies Press, 2008.

Wendt, Larry. "Sound Poetry: I. History of Electro-Acoustic Approaches II. Connections to Advanced Electronic Technologies." *Leonardo* 18, no. 1 (1985): 11–23.
Wendt, Larry. "Vocal Neighborhoods: A Walk through the Post-Sound Poetry Landscape." *Leonardo Music Journal* 3 (1993): 65–71.
White, Michael. "The Effect of the Nature of the Surround on the Perceived Lightness of Grey Bars within Square-Wave Test Gratings." *Perception* 10, no. 2 (April 1, 1981): 215–230.
Wiener, Philip P. *Dictionary of the History of Ideas: Studies of Selected Pivotal Ideas*. New York: Charles Scribner, 1973.
Wier, Dennis R. *Trance: From Magic to Technology*. Ann Arbor, MI: Trans Media, 2006.
Wier, Dennis R. *The Way of the Trance*. New York: Strategic Book, 2009.
Wiese, Wanja, and Thomas Metzinger. "Vanilla PP for Philosophers a Primer on Predictive Processing." In *Philosophy and Predictive Processing*. Edited by T. Metzinger and W. Wiese, 1–18. Frankfurt am Main: Mainz Johannes Gutenberg-Universität, 2017.
Wishart, Trevor. *On Sonic Art*. Amsterdam: Harwood, 1996.
Wordsworth, William, and Edward Moxon. *The Prelude, or, Growth of a Poet's Mind: An Autobiographical Poem*. London: E. Moxon, 1851.
Wright, Sylvia. "The Death of Lady Mondegreen." *Harpers* 11 (November 1954): 48–51.
Wyllie, Thomas R., and Adam Parfrey. *Love, Sex, Fear, Death: The Untold Story of the Process Church of the Final Judgement*. Los Angeles: Feral House, 2009.
Xenakis, Iannis, and Balint Andras. Varga. *Conversations with Iannis Xenakis*. London: Faber and Faber, 1996.
Young, La Monte. "Lecture 1960." *The Tulane Drama Review* 10, no. 2 (1965): 73–83.
Young, La Monte, and Marian Zazeela. *Selected Writings*. Edited by Michael H. Tencer. New York: Ubu Classics, 2004.
Youngblood, Gene. *Expanded Cinema*. New York: Dutton, 1970.
Z'ev. *1968–1990: One Foot In The Grave*. CD. Touch – TO:13, 1991.

Index

10 + 2 = 12 (American Sound Pieces) 51
10 Hours of Darth Vader Breathing 55
2001: A Space Odyssey (Kubrick) 67

Absolutego (Boris) 58
'Absolute Music,' aesthetic program 10
abstract sounds 172
accents, superimposition/alternation 149–50
accentuation
　compositional phantasmatic strategy/technique 79
　harp sonic phantoms 102
'access principle' 56
acousmatic music 9
acoustic event 32–3
acoustic meanings 159
acoustic niches, theory 184
acoustic synchrony 187
　collective displays 178–9
active concentration 102
Adding Machine, The 139
aesthetic content, introduction 60
aesthetic listening 101–2
akadinda (xylophone music) 69, 94–5
Akira (movie) 72
Alcock, James E. 133, 137
al-Haytham, Ḥasan Ibn 4
All Known Metal Bands (Nelson) 58
alpha rhythm stimulation 129
al-quat' (cutting process) 70
alternation, types (basis) 152
Amacher, Maryanne 47, 62, 69, 195, 197–8
amadinda (xylophone music) 69, 94–6, 155
ambient intelligence 31
ambiguity
　creativity, combination 31
　intentional practice 38
American Federation of the Arts conference (1957) 123

American Journal of the Society for Psychic Research, The (findings) 136
Amirkhanian, Charles 51, 165, 166
amplification, 'close-mic' style 89
Analog # 1 (Noise Study) (Tenney) 76
An Archivist's Nightmare (RLW) 168
Andriessen, Louis 74
Angel Moves Too Fast to See, An (Chatham) 35
animal calls
　'beauty,' appreciation 182
　micro-timbral structure, spectrogram 188f
anklung (eight-pitched bamboo) 155–6
Apollinaire, Guillaume 115
apophenia 11, 140
　creativity, relationship 22
　understanding 19–20
apophenic dynamics, description (example) 1
Arnheim, Rudolf 13, 123
Arnold, Martin 46, 159
Arnulf Rainer (Kubelka) 129
Arom, Simha 73
Arp, Hans 41, 42
Art and Visual Perception (Arnheim) 13
artification 91
　behaviors 91
Asante *ntahera* trumpet ensembles 72
Ashby, W. Ross 16
Ashley, Robert 77, 118
Assignment No. 1: Copying the "Orchid Pavilion Preface" 1,000 Times (Zhijie) 168
Atmosphères (Ligeti) 66
attention-independent processing 33
audible information, deficit 141–2

audio bistability 137
audio drumming 128
audio fidelity, *welcoming* degradations 142
audio images, creation 180
audio multi-tracking software, usage 57–8
audio-phantasmatic 167
auditory apophenia 30, 137
auditory driving (sonic entrainment) 48, 128–30
auditory fictions 10
auditory hallucinations 30
auditory illusions 30
auditory pareidolia 30
auditory perception 95
auditory scene analysis (ASA) 32, 36
  problem, introduction 33
Auditory Scene Analysis (Bregman) 96, 101, 142
auditory streaming 96
auditory streams 10, 32, 98
  segregation 174–5
auditory sub-streams 174–5
aural 'meaningful' phantasmatic patterns 182
authenticity, valuation 16
'automated drawing' sessions, graphic results 119f
automatic drawing, ghosts/dissociation 113
Automatic Writing (Ashley) 118
automatic writing technique, legacy 117–18
'avant' tendencies, injection 57–8

Ball, Hugo 124, 164
Banks, Joe 141
Bateson, Gregory 31, 59, 77
Bayesian brain frameworks 22
Bayes, Thomas 16
beatboxing 74, 156–7
'Beat Hotel' 138
beat-producing machines, usage 156
Becker, Judith 52, 125
'beholder's share', neurological phenomenon 24
belongingness, *Gestalt* rule 100
Berendt, Joachim-Ernst 56
'Berenice' (Poe) 167

Berlioz, portrait (example) 23f
'better-safe-than-sorry' strategy 20
bilateral action 123
bioacoustics, understanding 184
bipolar disorder 29
Birdlike (Hausmann) 154
bistability
  example 21f
  understanding 37
bistable pareidolic images, examples 23f
Black Sabbath 58
Black, Sheila 170
Blanchot, Maurice 54
Blom, Jan Dirk 28, 29
Blonk, Jaap 153
blotscapes
  examples 42f
  sonic blotscapes, instrument preparation (relationship) 87
Blotto (klecksographie) 41
Blu-Tack adhesive material, usage 86
Boethius, Anicius M.S. 80, 195
Bohor (Xenakis) 68
Bohr, Niels 114
Bol music 156
bolna ('to speak') 156
Bol Processor 156
Book of Love Being Written as They Touched (Weisser) 66
Bòorii adherents 126
Boris 58
Bos, Abbé du 177, 192
'boundary loss' 112
bracing 197
Branca, Glenn 35
Breakthrough (Jürgenson/Raudive) 136, 139
Bregman, Albert 32, 96, 101
Breton, André 115, 116
Brillopad That Looks Like Ireland, The (Hanafin) 15f
Broadsheet Book of Unspecified Things That Look Like Ireland, The 13
brown noise 133
Brugger, Peter 26
Buganda, court composers 95
Buñuel, Luis 117
Burnham, Jack 129
Burroughs, William 138–9

Index

Cage, John 76, 78, 117
calligraphy (Chinese characters) 168
cantu a tenòre 148
Can You Survive 10 Hours of Patrick Star Asking "Who You calling Pinhead" 55
Cardamom Mountains rainforest (Cambodia), nocturnal sonic environment 185f
Carroll, Peter 12
Castillo, Richard J. 127
Center for Applied Hearing Research 197
'chance image,' auditory versions 42
Changizi, Marc 192
Chatham, Rhys 35
Cherry, Colin 33
chöd (Buddhist practice) 115
Chopin, Henri 53, 164–5
chorusing effect 113
Clapping Music (Reich) 66
Clark, Andy 3–4, 6–7, 27–8, 123, 134
close-up microphones, usage 180
Cloud That Looks Like Ireland, The (Donovan-Witt, Lynn) 2f
cochlea, spontaneous mechanical oscillations 198
cognitive processing, 'top-down' model 24
Cohen, John 11
collective performance, drawing process (prototypical 'traces') 110f
Colley, Joe 181
Collins, Nicolas 172
Come to Free the Words (Gysin) 54
Come Out (Reich) 65
Composition 1960, #5 (Young) 61
compositional phantasmatic strategies/techniques 45
    compositional illumination 102
    accentuation 79
    layering 65
    noise 74
    persistence 55
    repetition 45
compositional series
    Drawing Phantoms 107
    Harp Phantoms 85
    Natural Phantoms 177
    Vocal Phantoms 147

'conceptual writing' project 168
conductus (music form) 70
Conrad, Klaus 12
Conrad, Tony 62–3, 129
Consolation of Philosophy, The (Boethius) 80
constant background broadband noise 30
contact mics (piezo-electric microphones), usage 89, 180
    amplification 98
Continuum (Ligeti) 67, 74
controlled hallucination 27
    perceptual phantoms state 8
courante (Schaeffer) 49
Cozens, Alexander 41
creativity
    ambiguity, combination 31
    apophenia, relationship 22
Crickets (Saroyan) 51
Croiners (Levine) 46
Crosby, Bing 134
crossdephasing 149
'cultural contexts' 20
'cut-up' methodology 138
cybernetics 124
    basis 114
CyberSongs 159
    compositions 160
    live performance, real-time video projection 160f
    musicalization 172
    suite, techno-cognitive-musical link 162
    text-to-speech-to-song pieces 158

Dalglish, William 70
Dalí, Salvador 117
Dau, Torsten 197
Davies, Rhodri 86, 88
da Vinci, Leonardo 41
Dead Pigeon on a Branch That Looks Like Ireland, The (O'Halloran) 15f
De Arithmetica 80
deceit, intentional means 178
De consolatione philosophiae (Boethius) 80
De Institutione Musica (Boethius) 80, 81
Denham, A.E. 31

Der Zauberlehrling (Ligeti) 73
Details Agrandis ('Enlarged Details')
    (Günter) 118, 141
Deutsch, Diana 53, 159
Devil's Music 1 (Collins) 172
Diamorphoses (Xenakis) 67
digitally generated blotscapes, examples 42f
Dissanayake, Ellen 91
dissociation (automatic drawing) 113
distortion-product otoacoustic emissions
    (DPOAEs) 197
Dixon, Deborah 139–40
DIY human synthesizer 158
Docta Sanctorum Patrum (Pope John
    XXII) 71
Dokaka 158
Dolden, Paul 66
dovetail interlocking techniques, usage 86
DPA omnidirectional mics, usage 104
drawing
    'automated drawing' sessions, graphic
        results 119f
    automatic drawing, ghosts/dissociation
        113
    drawing-photograph composite 143f
    EVP 108
    process, prototypical 'traces' 110f
    sounding 107
Drawing EVP 135
Drawing Phantoms 45, 64, 75, 82
    compositional series 107, 161
    graphic result, usage 143f
    live performances, real-time video
        projection (usage) 122f
    performance 121, 130, 144
    program text descriptions 135
    repetitive motor activities, operation
        126–7
    stage scenery, details 131
Drawing Phantoms séance (Ellison) 106f,
    131
    live performance 132f
Drawing Room, The 109
    collective performance, drawing
        process (prototypical 'traces') 110f
    collective performance, performance
        space 111f
    performance 120

Dream House, The 61
Dream Machine (Gysin) 68
'Dress, The' 23–4
drum dance, song 148
drumming, prolongation 128
Dufrêne, François 164–5
dynamic patterning 158

Earth 2: Special Low Frequency Version
    (Black Sabbath) 58
Eastern mysticism, mythical rigor 63
'ecology,' definition (expansion) 180
Écriture Automatique (Günter) 118
ecstatic trance 125
Edison, Thomas 140
ek stasis 126
ekstasis 129
Electronic Dance Music (EDM), techno
    (confusion) 169
electronic feedback 142
electronic music, hauntological genre 143
electronic sampling 157
Electronic Voice Phenomena (EVP) 46,
    123, 135
    defense 140
    electronic elicitation 144
    ITC treatment 138
    OVP, contrast 144
    prototypical arguments 137
    recordings 107
    research 139–40
    site-specific variant 136–7
    sonic phantoms study, association
        192–3
    technical/procedural aspects 141
    techniques 46
    voices, discovery 137
Eliade, Mircea 40
em-ha (vocal unit) 151
empress 'trumpet' cicada *(Megapomponia
    imperatoria)*, repetitive calls 189
endurance, effects 55
energy profile 158
Eno, Brian 63–4
environmental recordings, fieldwork
    location 176f, 179f
Esposito, Michael 139
events, horizontal chain 59

everyday trance 123
EVOL (Sonic Youth) 50
expanding/narrowing 123
'expansion through reduction,' presence 166
experiences, commonality 23
external synchronization, process 113
extrasensory perception, usage 144

face-seeing 20
Fäglar i Sverige (Fahlström) 165
Fahlström, Öyvind 165
Fat Boys 157
feedback process 92
feed-forward process 16
Feldman, Morton 56
Fellini, Frederico 72
Fem (Ligeti) 73
field recordings 181, 188, 191
fieldwork location
    Great Otway National Park 176f
    rainforest (Borneo, Malaysia) 179f
figure-and-ground focus points 130
'figure and ground' illusion 36
'figure and ground' perceptual situation 92
'figure and ground' relationship 69
'figure/ground' perceptual situation, manipulation 79
'first order superpositions' 195
Five Pieces for Orchestra (Schoenberg) 73
Flicker, The (Conrad) 129
Fludd, Robert 80
Focal Point (MacOS) 183
formant tracking 154
fortissimo broadband sonic broadcast 189
Fresh, Doug E. 157
Friston, Karl J. 25
Frith, Chris 6, 22
frog/insect sounds, interlocking patterns (spectrogram) 185f
From Here to Infinity (Ranaldo) 49
Frosch, Franz 198

gamelan mulut (mouth gamelan), usage 72
gamelan, usage 156
Ganzfeld experiments 60, 167
Ganzfeld states, simulations 142

Geiersbach, Frederick 57
Gendreau, Michael 48, 170
Gennep, Arnold van 39
Gestalt 69, 152, 182
    belonging 98
    belongingness rule 100
    Gestalt-psychological phenomenon 95
    Law of Good Continuation 35
    laws 99
    organizational laws 97
    psychological phenomenon 95
    psychology, application 13
    sequence, sub-streams 99
    similarity/proximity laws 99, 102
    theories 60
    voices/sounds, combination 69
ghost voices 140
ghosts (automatic drawing) 113
Gillis, Manon Anne 46
Giorno Poetry Systems 51
glissandi (gestures) 86
'global optimized performance' 135
Gnazzo, Anthony 51
'God of the in-between' 39–40
Goldsmith, Kenneth 168
Gombrich, Ernst 17, 25
gong kebyar (gamelan music) 72
Gordon, Michael 74
'grace,' condition 177
Gramaphone, Film, Typewriter (Kittler) 75
granular synthesis microsound 163–4
graphic scores, prevalence (increase) 187
Grauer, Victor 156
Great Otway National Park, fieldwork location 176f
Gregory, Richard 20
Grief (Throbbing Gristle) 139
Group Ongaku 117
Grush, Rick 5
guitar plectrum, usage 90
Günter, Bernhard 118
Gurney, Edmund 144
Gwiazda, Henry 183
Gysin, Brion 53–5, 68

habituation, effects 55
hallucinations
    controlled hallucination 27

controlled hallucination, perceptual
    phantoms state 8
definitions 28
human brain generation 6
non-psychiatric auditory
    hallucinations 29–30
uncontrolled perception 7
ha-mahe-ma (vocal unit) 151
Hanson, Sten 50
'happy genius' 26
Haraway, Donna 180
harmonic ratios 97
Harmonie Universelle (Mersenne) 81
harmonizing notes, absence 71–2
Harnessed (Changizi) 192
Harp Phantoms 45, 64, 75, 84f, 85
    alternating notes 103
    auditory streaming 96
    compositional series 85–7, 96, 130, 161
    dynamics, alterations 102
    inherent patterns 94
    listening modes, perceptual
        competition (relationship) 101
    looping process 162
    performing, approaches 94
    pieces 98, 104
    pitch/amplitude comparison/contrast 98
    project 82
    score, function 93
    second life, studio (usage) 104
    sonic figure/ground 100
    structure/variations 93
    three-note patterns 100
    thumb/finger, alternation 99
    workshop sessions 87f
Harp Phantoms performance
    reductions 94
    ritualization 91
Harp Phantoms score
    components 88
    function 93
    portion 93f
Harp preparation, diagram/photographic
    illustrations 89f
harp sonic phantoms, accentuation 102
harp sound, micro-element 103
hauntological genre 143
Hausmann, Raoul 154, 164

Hausswolff, Carl Michael von 139
Head Rhythm 1/Plaything 2 (Amacher) 197
hearing
    inherent/subjective patterns 10
    loss, complaints 29
Hebb, Donald 48
Hegarty, Paul 77
Heidsieck, Bernard 166
Helmholtz, Hermann von 4, 25, 78
Hine, Phil 55
hip-hop, Golden Age 157
hi-yah (vocal unit) 151
hochier (to shake) 70
Hocket (Monk) 74
hocket, impact 71
hocketing 156
Hohwy, Jakob 4, 6, 22
Hoketus (Andriessen) 74
Höller, Carsten 97
hollow mask illusion 20
Holy Trinity, indivisibility 114
hoquetus/hoquet (hiccup/stutter) 70
Hudak, John 180
Huggins pitch illusion 77–8
human perception, pattern recognition
    (impact) 12
*Human Personality and its Survival After
    Bodily Death* (Myers) 116
human universality 193
human voice, versatility 164
Huxley, Aldous 177
Hyde, Lewis 40, 42
hydrophones, usage 180
hyperpriors, determination/constraint 22
hyper-repetition, usage 172

ideomotor effect 120
Ikeda, Ryoji 68
illumination 102
illusions 5
    ancient/recent vocal techniques 153
IMAX theaters, usage 181
in-between ambiguous spaces 131
in-between moments 130
induced trance 123
informational meanings 159
inhaled/exhaled sounds, alternating
    patterns 155

inherent patterns 94
initial material explorations 107
inkblot psychological test (Rorschach) 41
insomnia, complaints 29
Instrumental Phantoms 68
instrumental sounds, mechanical-sounding loops 130
instruments
   preparation, sonic blotscapes (relationship) 87
   realm 85
   sonic exploration 85
   structure/variations 93
interlocking
   mesmeric effect 72
   natural polyphony 184
   patterns, frog/insect sounds (spectrogram) 185f
   patterns, impact 96
intermediate feedback, perception 150
intermodulation 198
interplanetary communications 140
'interplay,' process 69
intonation contours 151
Iqaluit (Baffin Island, Nunavut) landscape 146f
IRL 31
It Belongs to the Cucumbers (Burroughs) 139
ITC. *See* World Instrumental Transcommunication
It's Gonna Rain (Reich) 65

James, William 14
Janet, Pierre 114
Jastrow, Joseph 19
John Somebody (Johnson) 52
Johnson, Mimi 118
Johnson, Scott 52, 66
Joseph, Brandon 62
Jürgenson, Friedrich 136

kanashibari 22
Kandel, Eric 17, 25
Kant, Immanuel 4
   Kantian element 4
Karelian singing 128

Karetak, Vinnie 147
katajjaq 74, 82, 96, 147
   connection 155–6
   cultural preservation 149
   inventiveness/playfulness 158
   non-syllabic patterns/inhaled sounds 157
   performers 146f
   throat games 151
   vocalization 154
Kato, Masaharu 14
kecak (monkey chant technique) 72–3
Keeping Together in Time (McNeill) 112
Keller, Alex 58, 59, 140
Kemp, David 195
Kemp tones 195
khargyraa 147
khoomei 147
Kick That Habit, Man (Gysin) 54
'kinetic sculpture' (Gysin) 68
Kirkegaard, Jacob 197–8
Kittler, Friedrich 75
Klangfarbenmelodie (musical technique) 73
knowledge-driven model 4
Kokomo hum 196
'kosmiche' (bands) 63
Kosugi, Takehisa 117
Kraft Ainu, *rekkukara* 150
Krause, Bernie 63, 183–4, 192
Krauss, Rosalind 130
Krimanchuli (vocal polyphony) 73
Kris, Ernst 17
Kubelka, Peter 129
Kubik, Gerhard 10, 73, 94–5
Kubrick, Stanley 66–7

Labyrinthitis (Kirkegaard) 197
Lacy, Steve 54
La Légende d'Eer (Xenakis) 67
Lamb, Alan 61
'language poetry' 168
'Law of Good Continuation' 35
layering
   compositional phantasmatic strategy/technique 65
   natural polyphony 183
Layman's Guide to Cod Surrealism, A (Nurse With Wound) 118
Lecture on Nothing (Cage) 76

Led Zeppelin 158
Les Automatistes 117
Les Mandibules du Déjeuner sur l'Herbe ('Mandibles of Luncheon on the Grass') (Chopin) 165
Lessard, Ron 50
Lé Ventre de Bertini ('Bertini's Stomach') (Chopin) 165
Lewis-Williams, David 19, 20, 128
Licht', a 29-hour Cycle of Seven Operas (Stockhausen) 56
Ligeti, György 61, 66, 73–4
Lilly, John C. 173–4
listening 177
  level 102
  modes, perceptual competition (relationship) 101
  primitive mode 174
literal 'phono-graphic' performance, initial material explorations 107
live performance, real-time video projection (usage) 160f
live voices, *Vocal Phantoms #18* (usage) 151
L'Ivresse de la Vitesse (Dolden) 66
locked-grooves records 46
logarithmic connection 68
looping
  cycles 182
  ground pattern, establishment 92
  process 162
  technique, ground rules 163
looping-groove technique 49
looping/repetition strategy 48
loop multiplicity 130
López, Francisco 36, 64, 81, 83, 108, 177–8, 181
Lorente de Nó, Rafael 48
losing oneself 112
Lucier, Alvin 57, 117, 124
Lux Aeterna (Ligeti) 61

Machen, Arthur 114
MacKay, Donald 59
'magic spells' 111
Malavasi, Rachele 184
Malleus Maleficarum 141

Manely Brown, Rahzel 157
Manousakis, Stelios 147
man, portrait (example) 23f
Manyo'shu (Japanese poetry) 75
Marsching, Jane D. 24
Martínez, Israel 181
Masson, André 116
Massumi, Brian 28
materials
  explorations 107
  structural organization, alternation types (basis) 152
Matsubara, Sachiko (Sachiko M.) 56, 57
Matta, Ramuntcho 54
Mavromatis, Andreas 20
mbira (thumb pianos) 75
McCombe, Christine 180
McNeill, William 112
m-components 170
megaphone (Nattiez) 150
mental fatiguing effect 158
Mersenne, Marin 81
Metallica 158
Metaphoric Postcard 23f
micro-particles 165
micro-temporal mechanisms 162
micro-timbral component, audibility 189
micro-TTS loop units 161
microsound, granular synthesis 163–4
migraine, complaints 29
mimesis, vocal techniques 154
Missing Sense, A (Nurse With Wound) 118
Mizuno, Shuko 117
Mmabolela Reserve (South Africa)
  fieldwork locations 186f
  nocturnal recording, spectrogram 186f
Moles, Abraham 59–60
Monk, Meredith 74
monomania 169
  pieces 161
  psychological state 167
Moore, Christopher G. 22
Mori, Ikue 86
morpheme patterns 151
motet (music form) 70
moths, jamming talents (outwitting) 178

motifs 151
    combination 152
    organization, example 153f
    performing, alternation 152
    unison configuration 152
MTV 125
Mugitani, Ryoko 14
multi-channel arrays, usage 180
multifunctional host-structure 148–9
multi-layered rainforest dusk transition
        (Borneo, Malaysia), spectrogram 190f
Muses, Madmen and Prophets (Smith) 28
Mushrooms (for John Cage)
        (Amirkhanian) 51
Musica Humana 81, 82
Musica Iconologos (Tone) 75
Musica Instrumentalis 81, 82
musicalization 158–9, 172
    process 159–60
Musical Offering, The (Bach) 73
Musica Mundana (Music of the Spheres)
        81, 82
Musica Simulacra 75
Musica Universalis 81
music, categories 81
Music for Solo Performer (Lucier) 124–5
music, temporality 178
musique concrète 9, 35–6
My First Motorola (Nemerov) 168
Myers, Frederic W. H. 116, 144

naked woman, image (example) 23f
Namtchylak, Sainkho 147
Nattiez, Jean-Jacques 39, 148–9, 151, 154
Natural Phantoms 68, 178, 191
    compositional series 177
    compositions, inspiration 191
    compositions, sound processing 189
    inspiration 187
    series, compilation process 191
Natural Phantoms #1 191
Natural Phantoms #9 192
Natural Phantoms #10 192
natural polyphony 182
    interlocking 184
    layering 183
    transitions 189

natural world, 'audio images' (creation) 180
'nature cultures' 180
nature, realm 177
Nauman, Bruce 51
Nechvatal, Joseph 114
'Necker cube' 18f, 19
Necker, Louis Albert 18f
Nelson, Dan 58
Nemerov, Alexandra 168
Newton, Isaac 5
Niblock, Phill 56, 64
Nicolai, Carsten 68
Ninth Sonata (Beethoven) 60
Nketia, J.H. Kwabena 71–2
nocturnal sonic environment, Cardamom
        Mountains rainforest (Cambodia)
        185f
noise
    compositional phantasmatic strategy/
        technique 74
    compositions 76
    constant background broadband noise
        30
    floor, raising 143f
    listening experience 76
    noise-as-art/noise-as-music 75
    types 133
    white surface noise, inclusion 142
NON (Rice) 50
non-cognitive collaborators 181
No No No No (Nauman) 51
non-psychiatric auditory hallucinations
        29–30
Nó, Rafael Lorente de 48
normal listening experience 127
notation symbols, usage 152
N:O:T:H:I:N:G (Sharits) 129
'nude hidden with a diaphanous veil'
        (example) 97
Nurse With Wound 118
Nur Sur 148

object
    appearance, vividness 13
    realm 107
objective reality, experience 5
object-oriented culture 129

'Object Voice Phenomena' (OVP) 108, 135, 143
   EVP, contrast 144
   sonic phantoms study, association 192–3
O'Callaghan, Casey 8, 33, 34
off-beats 155
Oliveros, Pauline 61, 63
one-dimensional objects, appearance 13
On Intelligence (Taine) 29
on-site listening, fieldwork location 176f, 179f
Oomoonoon: Dancing on the Brink of the Word #1 (Weisser) 66
order/disorder, transitioning (experience) 105
order-seeking 51
organization, vertical modes 65
organum (music form) 70
'Original Human Beatbox' (Fresh) 157
'oscillation' 70
ostinatos
   figures 52
   two-note *ostinatos* (usage) 86
   usage 86
ostranenie (aesthetic theory) 171
Oswald, John 47
Other Voice, The 135
otoacoustic emissions (OAEs) 195
   discovery 196
   Phantom Within, The 195
Oval 47
overtone singing (Xhosa) 148
Owen Drumm piezo mic mixer, usage 104

Pagan Muzak (NON) 50
Palestine, Charlemagne 74, 197
Pan, Stephanie 147
paranoia 26
paranormal audio research 107
parapsychology, literature 144
Parataxes (Gendreau) 48, 170
pareidolia
   auditory pareidolia 30
   example 2f, 21f
pareidolic perception, adaptability 26

Parkins, Zeena 85
Partita for solo flute in A Minor (Bach) 73
Pater, Walter 10
pattern recognition, impact 12
Patterson, Lee 181
Penderecki, Krzystof 66
perceptions
   fantasies 6
   uncontrolled perception 27
perceptual bistable images, examples 18f
perceptual competition, listening modes (relationship) 101
perceptual phantoms 5
perceptual spotlight 91
perceptual systems, theory 135
performance
   collective performance, drawing process (prototypical 'traces') 110f
   live performances, real-time video projection (usage) 122f
   ritualization 91
   space 111f
periodic amplitude modulation 188
Perloff, Marjorie 168
'permutation poem' 54
Perotinus 70
persistence (compositional phantasmatic strategy/technique) 55
phantasmagenics 3
Phantasma Humana 74, 82
   CyberSongs (text-to-speech-to-song pieces) 158
   illusion, ancient/recent vocal techniques 153
   katajjaq 147
   micro-temporal mechanisms 162
   realm of the voice 147
   semantic satiation/semantization 167
   Vocal Phantoms #18 (live voices) 151
Phantasma Instrumentalis 82
   auditory streaming 96
   harp sonic phantoms, accentuation 102
   inherent patterns 94
   instrument preparation 87
   instrument, realm 85
   listening modes, perceptual competition (relationship) 101

performance, ritualization 91
sonic blotscapes 87
sonic figure/ground 100
structure/variations 93
Phantasma Materialis 82
   automatic drawing, ghosts/dissociation 113
   Drawing Room, The 109
   expanding/narrowing 123
   initial material explorations 107
   loop multiplicity 130
   realm of the object 107
   transcendent 'boundary loss' 112
Phantasma Naturalis 178
   composition, expansion 192
   interlocking 184
   layering 183
   listening/recording 177
   Natural Phantoms 83
   Natural Phantoms (compositional series) 177, 191
   natural polyphony 182
   nature, realm 177
   transitions 189
phantasmatic
   compositional phantasmatic strategies/techniques 45
   induction 45
   sonic phantasmatic experience, realms 80
phantasmatic emergent phenomenon 8
phantasmatic transmitters 198
phantom-inducing power 161
phantom patterns, asymmetrical shape 152
Phantom Plastics (record label) 139
Phantom Portrait 12f
Phantasms of the Living (Myers) 144
Phantom Within, The 194f
   otoacoustic emissions 195
'phantom word illusion' 54
Phantom Words and Other Curiosities (Deutsch) 53
phantom words illusion 162
phonemes, perception 159
phono-graphic description 107
'phono-graphic' performance, initial material explorations 107
photic driving 129

pič Eynen (recordings) 148, 150–1
piezo-electric microphones (contact mics), usage 89, 180
   amplification 98
pileojartuq (two-person contests) 148
pink noise 133
Pissarro, Camille 17
Pistol Poem (Gysin) 54
pitch/frequency characteristics, presence 198
Plantenga, Bart 74
Platz, Robert 36, 38
Pléïades (Xenakis) 74
Podmore, Frank 144
Poe, Edgar Allan 167, 169
Poème Symphonique (for 100 Metronomes, 10 Performers and 1 Conductor) (Ligeti) 66
'poet to come' role (Spare) 117
pola cak chorus 73
Pollard, Cayce 22
Pollock, Jackson 117
polyglot speech 137
Polymoog, usage 118
polyphony 82
Pope John XXII 71
Popol Vuh 63
Population Explosion (Gnasso) 51
post-Freudian analysis 116
post hoc data analysis 19
post-humanities 180
post-traumatic stress disorder (PTSD) 29
Potebnya, Alexander 171
prediction engines 133
prediction errors, avoidance 11
predictive coding (PC) 4
predictive processing (PP) 4, 22
Prelude, The (Wordsworth) 1
presence, apophenia (relationship) 11
Principles of Music, The (Boethius) 80, 81
Probst, Rudolf 196
prototypical 'traces' 110f
proximity influence, *Gestalt* laws 99, 102
pseudo-polyphony, generation 34
psychedelic music 58
psychological doubleness 74
psycho-physical tests 76

psychosis 26
Puddi Puddi 10 Hours 55
puirt à beul (mouth music), origins 156
Punktefilm (Höller) 97
pure electronic tones, high-frequency hocket 197
pure phonism 165
pure tones 35

quadrophonic sound system, usage 109
quantum sonics 164

'Rabbit and Duck' image 18f, 19, 36
Radigue, Eliane 56, 63, 64, 197
radio frequency spectrum, partitioning 183
'Radio Peter' 137
Rahzel 157
rainforest
    Cardamom Mountains (Cambodia) nocturnal sonic environment 185f
    fieldwork location (Borneo, Malaysia) 179f
    multi-layered rainforest dusk transition (Borneo, Malaysia) spectrogram 190f
Ranaldo, Lee 49–50
Raster/Noton (label) 68
Raudive, Konstantin 136, 137, 140
Raumform (Platz) 36
Raw Materials (Nauman) 51
Ray Gun Virus (Sharits) 129
reality, distortion 9
real-time video projection, usage 122f, 160f
recording 177
    ambience 141
'reduced listening' 172
reflexive feedback loops 114
Reich, Steve 65–6
Rekukhara 148, 150
'repeating word effect' 173
repetition (compositional phantasmatic strategy/technique) 45
Representative List of the Intangible Cultural Heritage of Humanity 147–8

reverse engineering 17
rhythmic activity, shifts 105
rhythmic patterns 151
Rice, Boyd 50
Ricercar a 6 (Bach/Webern) 73
Riegl, Alois 16
ritualization 91, 92
RLW. *See* Wehoswky
Roads, Curtis 164, 166
Robindoré Brigitte 164
Roederer, Juan 195
Roland TR-808, usage 156
Rorschach blots 42
Rorschach, Hermann 41
Rosenboom, David 57
Rosenfeld, Edward 174
Rösterna från Rymden ('Voices from Space') (Jürgenson) 136
Rouget, Gilbert 129
Royal Courts (amadinda/akadinda xylophone music) 69
RRRecords 50
'Rubin goblet' 36

Sagan, Carl 14
Salvi, Angélica V. 86
    set-up tests 89
sampling, musical concept 35
Saroyan, Aram 51
Satie, Erik 47
saturation, effects 55
Satyricon (Fellini) 72
Scape (journal) 180
scene-analysis processes, defeat 34
scenes, selection/modular combination 94
Schaeffer, Pierre 9, 35, 49, 172
Schafer, R. Murray 187
Schapiro, Meyer 123–4
schizophonia 35–6
schizophrenia 29
Schoenberg, Arnold 73
Schopenhauer, Arthur 10
Schreiber, Klaus 138
Schwitters, Kurt 164
Scruton, Roger 10
séance

spiritualist séance 131
  voices 131, 133
self-inquiry, methods 124
semantic meanings 159
semantic satiation 167, 169–70
semantization 167
semi-automatic composition, process 191
sensory deprivation, perspective 30
Serres, Michel 142
Seth, Anil 5, 6, 17
Sharits, Paul 68–9, 129, 170
Shermer, Michael 19
Shih Ching (Chinese poetry) 75
Shklovksy, Viktor 171
'silent prayer' 27
similarity, *Gestalt* laws 99, 102
sine-wave speech 134
Skudučiai (flute music) 73
Smalley, Denis 158, 172
Smith, Daniel 28
Smith, William 60
Smoor, Nathalie 131
Snowy Town 43f
Something Bumped against a Wall at Work and Made a Painting of a Snowy Town 43f
Sommerville, Ian 138
Somnimage (record label) 64
Sonal Atoms (Roads) 164
Sonata for Recorder in C major (Telemann) 73
sonic adventure, ironies 49
sonic animism 182
Sonic Art (Wishart) 78
sonic blotscapes 140
  instrument preparation, relationship 87
  miniaturized sonic blotscapes 169
  sources 181–2
sonic creativity, evolution 39
sonic entrainment (auditory driving) 48, 128
sonic experience 38
sonic figure/ground 100
sonic illusion, reality 35
sonic overlays, impact 195
sonic phantasmatic experience, realms 80

sonic phantoms 5
  emergent presence 3
  generation 154–5
  harp sonic phantoms, accentuation 102
  inherent/subjective patterns 10
  sonic phantom-producing polyphony 95
Sonic Phantoms compositional project 178
sonic reality, unfolding 181
sonic tapestries 178
sonic transparency 31
Sonic Youth 50
soramimi ('ears in the sky') 14
Sorel, Georges 24
Sound Characters (Amacher) 197
sound materials 150
Sound Poems for an Era of Reduced Expectations (Wendt) 166
sound poems, impact 165
sounds
  events, timeline 186–7
  gesture/texture 158
  installation 173f
  processing, usage 189
  voices, combination 69
Spare, Austin Osman 114, 115, 117
Special Low Frequency Version (Boris) 58
'spectral glide' 154
spectrogram
  animal calls, micro-timbral structure (prototypical examples) 188f
  frog/insect sounds, interlocking patterns 185f
  multi-layered rainforest dusk transition (Borneo, Malaysia), prototypical example 190f
  nocturnal recording, Mmabolela Reserve (South Africa) 186f
speech-based compositions 52
speech-based exercises 52
speech sonic patterns, hyper-repetition (usage) 172
speech-to-song effect 158
speech-to-song illusion 159

speech-to-song phenomenon 160
spellbound, feeling 110
Spiricom 144
spiritualist séance 131, 136
spontaneous otoacoustic emissions (SOAEs) 197
SPR 144
Sprechfunk mit Vestorbenen (Voice Transmissions with the Deceased) (Jürgenson/Raudive) 136, 139
Starting from Maya (Wendt) 166
St. Bernard of Clairvaux 71
Stella, Frank 56
Stepputat, Kendra 72
Steps to an Ecology of Mind (Bateson) 77
stimulus-frequency otoacoustic emissions (SFOAEs) 197
Stitt, André 40
Storm of Drones (Asphodel) 64
straight field recordings 191
straight performance, approach 94
stream segregation
    processes 105
    theory 97
stress, complaints 29
stridulation 188
String Quartet No. 2 (Feldman) 56
Strogatz, Steven 113
structural timeline, example 93f
Strumming Music (Palestine) 74
studio, usage 104
Suède paréidolie (example) 21f
Sugar Fish Drink (Nurse With Wound) 118
superhuman behavior 128
super-psi 144
Surrealist movement 114
    group manifesto 115–16
Suzuki, Dean 50, 66
Swarm of Drones (Asphodel) 64
SwissBeatbox 74
synaptic strength, increase 48
synchronous chorusing 187
synchronously evoked otoacoustic emissions 197
synesthesia, abnormality 24
Szalay, Attilla von 136

Taine, Hippolyte 29
Tangerine Dream 63
Taos hum 196
Tashi Gomang (Oliveros) 63
techno, confusion 169
Telemann, Georg Philipp 73
temporal persistence, opposite 170
temporal relationships, perception 53
Tenney, James 76
Ter Veldhuis, Jacob 52
Tesla, Nikola 140
tessitura 52
text-to-speech (TTS) 160–1
    micro-TTS loop units 161
    pieces, making 161
    synthesizer, usage 166
text-to-speech computer technology 159
text-to-speech-to-song
    pieces 158
    usage 173f
Theatre of Eternal Music, The (Young) 61, 62
Theosophical Society 113
Think Think Think, OK OK OK, Work Work (Nauman) 51
Third Voice, The 44f
Thompson, Tok 156, 157
Thomson Virgil 117
three-note patterns, usage 100
Threnody (for the Victims of Hiroshima) for 52 Strings (Penderecki) 66
throat-singing variants 140
throat song 149
Throbbing Gristle, influence 138
Throne of Drones (Asphodel) 64
timbral complexity, enhancement 90
Timbre (Gordon) 74
Tone, Yasunao 75, 117
top-down processing 139
top-driven model (knowledge-driven model) 4
Totally Corrupt (Giorno Poetry Systems) 51
T,O,U,C,H,I,N,G (Sharits) 170
Tourette's syndrome, presence 118
'traces' 110f
trance

activity, afterglow 126
dissociation, experience 130
induction 55
institutionalized forms 127
trance-inducing drawing ritual 82
trance-like experiences, induction 112
trance-like state 109, 116–17
Trance Research Foundation 127
transcendent 'boundary loss' 112
transformation 129
transition phases 131
transitions (natural polyphony) 189
Treatise about Painting (da Vinci) 40
trickster archetype 114–15
Trickster Makes This World (Hyde) 40
Turner, Victor 39
Turrell, James 142
two-note *ostinatos* (usage) 86
tymbalization 188
Tzadik (record label) 197
Tzara, Tristan 41, 164

Umngqokolo 148
Umwelt 177
'uncontrolled perception' 27
Unhörbares wird Hörbar (The Inaudible Becomes Audible) (Jürgenson/Raudive) 136
universal truths, transmission 63
Un Peu de Neige Salie ('A Little Bit of Dirty Snow') (Günter) 141
untitled #272 (López) 64
untitled #305 [seven nights] (López) 64
updated belief, equation 16

Vaya con Dios ('God be with you') (Collins) 172
Veldhuis, Ter 66
Very Low Frequency (VLF) receiver 26
Vexations (Satie) 47, 58
Vingt-cinq poèmes (Tzara) 41
Viola, Bill 125
Violin Phase (Reich) 66
virtual reality (VR) headsets, usage 181
visual avatar, creation 173f
visual landscapes, composition 42

visual stimulation 128
Vitebsky, Piers 39, 125
vocal illusionism 154
Vocal Phantoms 46, 64, 82, 147
  compositional series 147
  performance poster 44f
  pieces, construction 163
  video projection, visual avatar (creation) 173f
Vocal Phantoms #18 159
  live performance set-up, score/diagram (section) 153f
  live voices 151
vocal sound, spectral characteristics 159
voice
  ghost voices 140
  hearing 127
  realm 147
  recognition 133
  sonic phantoms, multiplicity 144
  sound, combination 69
  term, usage 182
Vocal Phantoms #18 (live voices) 151
  voice-expectant mental state, intensification 135
voiced/un-voiced sounds, alternating patterns 155
voiced/voiceless patterns 151
'voice-expectant' mental state intensification 135
Voice Phantoms 68
'Voices from Space' (Jürgenson) 136
von Foerster, Heinz 3

Waco Girls 58
Wave Field Synthesis sound system 103
Webern, Anton 73
Wehoswky, Ralf (RLW) 168
Weisser, Stefan 66
Well-Tempered Clavier (Bach) 80
Well-Tuned Piano (Young) 61
Wendt, Larry 165–6
Wenzel, Martin 138
White Christmas (Crosby) 134
white noise 133
  generator, usage 136

'White's illusion' 37f
white surface noise, inclusion 142
Wide Out (Turrell) 142
Wier, Dennis R. 127
wilderness environments, natural polyphony 178
Wild Sanctuary (soundscape archive) 183–4
Wishart, Trevor 65, 78, 154, 167
Wolfman (Ashley) 77
word/phrase, writing 169
Wordsworth, William 1
world, hierarchical generative model 4

World Instrumental Transcommunication (ITC), formation 137

Xenakis, Iannis 58, 67–8, 80

Yodel-Ay-Ee-Oooo (Plantenga) 74
Yodel in Hi Fi (Plantenga) 74
Young, La Monte 58, 61–2, 64, 78

Zazeela, Marian 61
Z'ev 54, 66, 124
Zhijie, Qiu 168
Zorn, John 86